JACARANDA
GEOGRAPHY ALIVE 8

VICTORIAN CURRICULUM | SECOND EDITION

JUDY MRAZ

CATHY BEDSON

ALISTAIR PURSER

BENJAMIN ROOD

DENISE MILES

CONTRIBUTING AUTHORS

ALEX SCOTT | JILL PRICE | KATHRYN GIBSON

JEANA KRIEWALDT | ALEX ROSSIMEL

jacaranda
A Wiley Brand

Second edition published 2020 by
John Wiley & Sons Australia, Ltd
42 McDougall Street, Milton, Qld 4064

First edition published 2017

Typeset in 11/14 pt TimesLTStd

© John Wiley & Sons Australia, Ltd 2020

The moral rights of the authors have been asserted.

ISBN: 978-0-7303-7950-8

Front cover image: © Claudiad/Getty Images Australia

Illustrated by various artists, diacriTech and Wiley Composition Services

Typeset in India by diacriTech

Printed in Singapore by
Markono Print Media Pte Ltd

A catalogue record for this book is available from the National Library of Australia

10 9 8 7 6 5 4

CONTENTS

HOW TO USE
the *Jacaranda Geography Alive* resource suite

The ever-popular *Jacaranda Geography Alive for the Victorian Curriculum* is available as a standalone Geography series or as part of the *Jacaranda Humanities Alive* series, which incorporates Geography, History, Civics and Citizenship, and Economics and Business in a 4-in-1 title. The series is available across a number of digital formats: learnON, eBookPLUS, eGuidePLUS, PDF and iPad app.

Skills development is integrated throughout, and explicitly targeted through SkillBuilders and a dedicated Geographical skills and concepts topic for each year level.

This suite of resources is designed to allow for differentiation, flexible teaching and multiple entry and exit points so teachers can *teach their class their way*.

Features

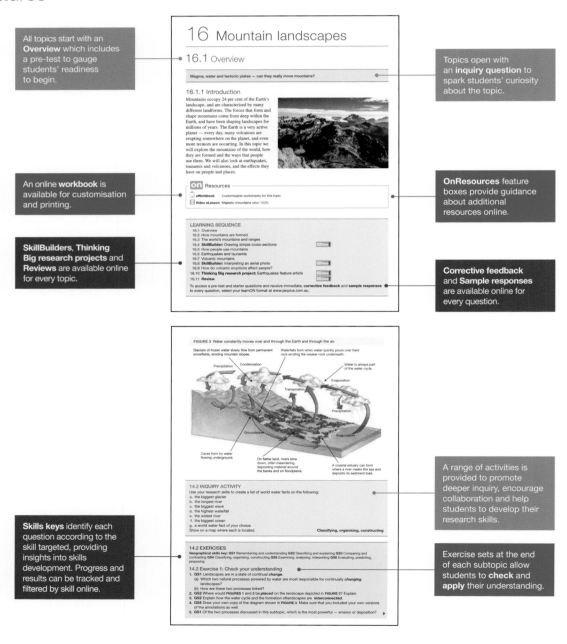

All topics start with an **Overview** which includes a pre-test to gauge students' readiness to begin.

An online **workbook** is available for customisation and printing.

SkillBuilders, Thinking Big research projects and **Reviews** are available online for every topic.

Skills keys identify each question according to the skill targeted, providing insights into skills development. Progress and results can be tracked and filtered by skill online.

Topics open with an **inquiry question** to spark students' curiosity about the topic.

OnResources feature boxes provide guidance about additional resources online.

Corrective feedback and **Sample responses** are available online for every question.

A range of activities is provided to promote deeper inquiry, encourage collaboration and help students to develop their research skills.

Exercise sets at the end of each subtopic allow students to **check** and **apply** their understanding.

Skillbuilders model and develop key skills in context.

Content is presented using age-appropriate language, and a wide range of engaging sources, diagrams and images support concept learning.

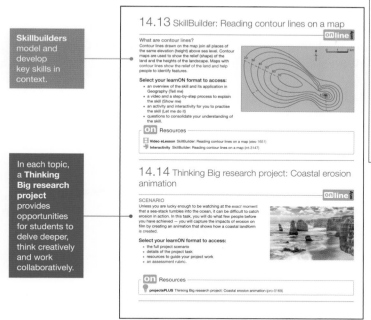

14.13 SkillBuilder: Reading contour lines on a map

What are contour lines?
Contour lines drawn on the map join all places of the same elevation (height) above sea level. Contour maps are used to show the relief (shape) of the land and the heights of the landscape. Maps with contour lines show the relief of the land and help people to identify features.

Select your learnON format to access:
- an overview of the skill and its application in Geography (Tell me)
- a video and a step-by-step process to explain the skill (Show me)
- an activity and interactivity for you to practise the skill (Let me do it)
- questions to consolidate your understanding of the skill.

On | Resources
- **Video eLesson** SkillBuilder: Reading contour lines on a map (eles-1651)
- **Interactivity** SkillBuilder: Reading contour lines on a map (int-3147)

In each topic, a **Thinking Big research project** provides opportunities for students to delve deeper, think creatively and work collaboratively.

14.14 Thinking Big research project: Coastal erosion animation

SCENARIO
Unless you are lucky enough to be watching at the exact moment that a sea-stack tumbles into the ocean, it can be difficult to catch erosion in action. In this task, you will do what few people before you have achieved — you will capture the impacts of erosion on film by creating an animation that shows how a coastal landform is created.

Select your learnON format to access:
- the full project scenario
- details of the project task
- resources to guide your project work
- an assessment rubric.

On | Resources
- **projectsPLUS** Thinking Big research project: Coastal erosion animation (pro-0169)

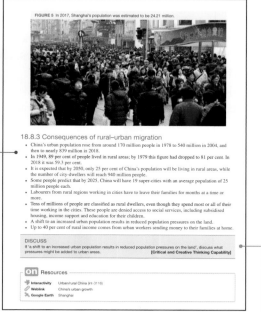

FIGURE 5 In 2017, Shanghai's population was estimated to be 24.21 million.

18.8.3 Consequences of rural–urban migration
- China's urban population rose from around 170 million people in 1978 to 540 million in 2004, and then to nearly 839 million in 2018.
- In 1949, 89 per cent of people lived in rural areas; by 1979 this figure had dropped to 81 per cent. In 2018 it was 59.3 per cent.
- It is expected that by 2050, only 25 per cent of China's population will be living in rural areas, while the number of city-dwellers will reach 940 million people.
- Some people predict that by 2025, China will have 19 super-cities with an average population of 25 million people each.
- Labourers from rural regions working in cities have to leave their families for months at a time or more.
- Tens of millions of people are classified as rural dwellers, even though they spend most or all of their time working in the cities. These people are denied access to social services, including subsidised housing, income support and education for their children.
- A shift to an increased urban population results in reduced population pressures on the land.
- Up to 40 per cent of rural income comes from urban workers sending money to their families at home.

DISCUSS
If 'a shift to an increased urban population results in reduced population pressures on the land', discuss what pressures might be added to urban areas. **[Critical and Creative Thinking Capability]**

On | Resources
- **Interactivity** Urban/rural China (int-3116)
- **Weblink** China's urban growth
- **Google Earth** Shanghai

Discuss features explicitly address Curriculum Capabilities.

Links to the **myWorld Atlas** and **myWorldHistory Atlas** are provided throughout.

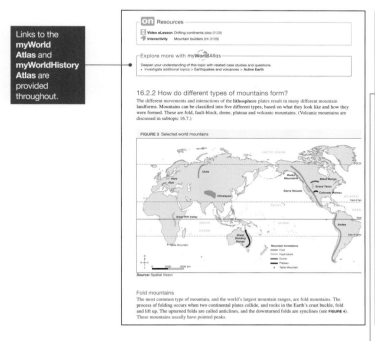

On | Resources
- **Video eLesson** Drifting continents (eles-0129)
- **Interactivity** Mountain builders (int-3109)

Explore more with myWorldAtlas
Deepen your understanding of this topic with related case studies and questions.
- Investigate additional topics > Earthquakes and volcanoes > Active Earth

16.2.2 How do different types of mountains form?
The different movements and interactions of the **lithosphere** plates result in many different mountain landforms. Mountains can be classified into five different types, based on what they look like and how they were formed. These are fold, fault-block, dome, plateau and volcanic mountains. (Volcanic mountains are discussed in subtopic 16.7.)

FIGURE 3 Selected world mountains

Source: Spatial Vision

Fold mountains
The most common type of mountain, and the world's largest mountain ranges, are fold mountains. The process of folding occurs when two continental plates collide, and rocks in the Earth's crust buckle, fold and lift up. The upturned folds are called anticlines, and the downturned folds are synclines (see **FIGURE 4**). These mountains usually have pointed peaks.

14.15 Review

14.15.1 Key knowledge summary
Use this dot point summary to review the content covered in this topic.
14.15.2 Reflection
Reflect on your learning using the activities and resources provided.

On | Resources
- **eWorkbook** Reflection (doc-31346)
  Crossword (doc-31347)
- **Interactivity** Landscapes formed by water crossword (int-7596)

KEY TERMS
avalanche a sudden downhill movement of material, especially snow and ice
backwash the movement of water from a broken wave as it runs down a beach returning to the ocean
barge a long flat-bottomed boat used for transporting goods
clinometer an instrument used for measuring the angle or elevation of slopes
deposition the laying down of material carried by rivers, wind, ice and ocean currents or waves
destructive wave a large powerful storm wave that has a strong backwash
downstream nearer the mouth of a river, or going in the same direction as the current
ecosystem an interconnected community of plants, animals and other organisms that depend on each other and on the non-living things in their environment
erosion the wearing away and removal of soil and rock by natural elements, such as wind and water, and by human activity
estuary the wide part of a river at the place where it joins the sea
field sketch a diagram with geographical features labelled or annotated
flash flood a flood that occurs very quickly, often without advance warning
floodplain an area of low-lying ground adjacent to a river, formed mainly of river sediments and subject to flooding
groundwater water that seeps into soil and gaps in rocks
hard engineering a coastal management technique that involves using physical structures to control the effects of natural processes
human features structures built by people
intermittent describes a stream that does not always flow
longshore drift a process by which material is moved along a beach in the same direction as the prevailing wind
meander a winding curve or bend in a river
moraine rocks of all shapes and sizes carried by a glacier
peninsula land jutting out into the sea
perennial describes a stream that flows all year
physical process continuing and naturally occurring actions such as wind and rain
prevailing wind the main direction from which the wind blows
river delta a landform created by deposition of sediment that is carried by a river as the flow leaves its mouth and enters slower-moving or stagnant water. Can take three main shapes: fan shaped, arrow shaped and bird-foot shaped
shell middens Indigenous archaeological sites where the debris associated with eating shellfish and similar foods has accumulated over time
soft engineering a coastal management technique where the natural environment is used to help reduce coastal erosion and river flooding
swash the movement of water in a wave as it breaks onto a beach
tributary a river or stream that flows into a larger river or lake

A range of questions and a post-test are available online to test students' understanding of the topic.

Key terms are available in every topic review.

learn**on**

Jacaranda Geography Alive learnON is an immersive digital learning platform that enables student and teacher connections, and tracks, monitors and reports progress for immediate insights into student learning and understanding.

It includes:

- a wide variety of embedded videos and interactivities
- questions that can be answered online, with sample responses and immediate, corrective feedback
- additional resources such as activities, an eWorkbook, worksheets, and more
- Thinking Big research projects
- SkillBuilders
- teachON, providing teachers with practical teaching advice, teacher-led videos and lesson plans.

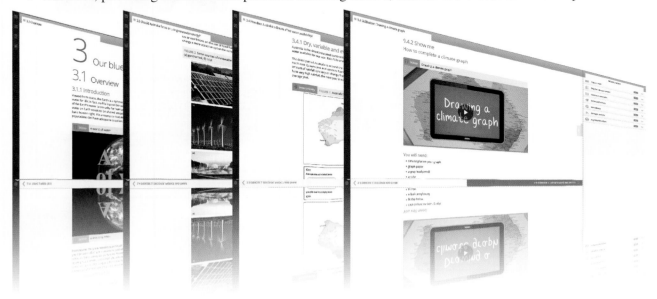

teach**on**

Conveniently situated within the learnON format, teachON includes practical teaching advice, teacher-led videos and lesson plans, designed to support, save time and provide inspiration for teachers.

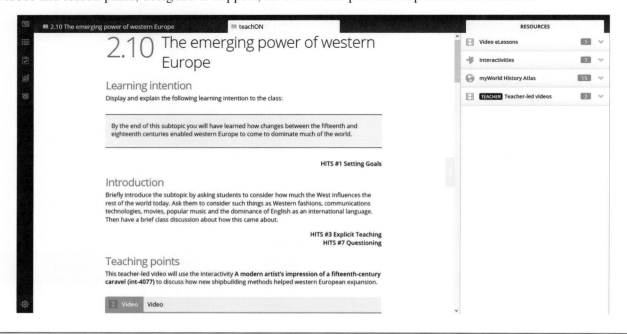

ACKNOWLEDGEMENTS

The authors and publisher would like to thank the following copyright holders, organisations and individuals for their assistance and for permission to reproduce copyright material in this book.

The Victorian Curriculum F–10 content elements are © VCAA, reproduced by permission. The VCAA does not endorse or make any warranties regarding this resource. The Victorian Curriculum F–10 and related content can be accessed directly at the VCAA website. VCAA does not endorse nor verify the accuracy of the information provided and accepts no responsibility for incomplete or inaccurate information. You can find the most up to date version of the Victorian Curriculum at http://victoriancurriculum.vcaa.vic.edu.au.

Images

• 123RF: **57** (bottom right)/Angelena Rebori • Alamy Australia Pty Ltd: **4** (bottom), **239** (top)/Pulsar Imagens; **9** (left), **55**, **93**, **98** (bottom)/National Geographic Image Collection; **19** (bottom), **270** (bottom left)/Global Warming Images; **27**/Mark A. Johnson; **30**/Bill Bachman; **61**, **165** (top right)/robertharding; **63** (bottom left)/© keith morris; **92** (top)/Ingo Oeland; **97**/Greatstock; **101** (top)/Accent Alaska.com; **121**/© redbrickstock.com; **128** (bottom)/DOD, S.Dupuis; **134**/epa european pressphoto agency b.v.; **150**/Amazon-Images; **164** (right)/David Tipling Photo Library; **168** (bottom)/Friedrich Stark; **170** (middle)/DEEPU SG; **173** (top)/Sue Cunningham Photographic; **183** (left)/Andrew Sole; **216** (f)/Mark Thomas; **232** (botttom right)/Universal Images Group North America LLC; **233** (bottom)/Kees Metselaar; **249**/David Wall; **251**/Chris Willson; **266**/Australia; **279** (top left)/Kirsty McLaren Stock Photo; **279** (top right)/ deadlyphoto.com Stock Photo; **280**/© TGB; • Alex Rossimel: **64** (top)/Alex Rossimel • Aussie Kanck: **252** (top), **253**/Aussie Kanck • Australian Antarctic Division: **103**/IA39475 Aerial view of Davis station, Vestfold Hills. Photograph © Darryl Seidel, courtesy Australian Antarctic Division • Australian Wildlife Conservancy: **43** (bottom)/Australian Wildlife Conservancy AWC/Wayne Lawler • Bhattacharyya and Werz: **217**/Bhattacharyya and Werz • Climate Council: **268** (top)/Climate Council, with data from American Council for an Energy-Efficient Economy ACEEE 2018 Scorecard • Coastal Studies Institute, LSU: **70**/Mike Blum & Harry H. Roberts/Coastal Studies Institute LSU Baton Rouge, LA • Creative Commons: **9** (right), **165** (bottom), **247** (top left), **262** (bottom left)/Creative Commons; **133** (bottom)/Domenico/flickr; **274** (bottom)/Jan Seifert • Department of Home Affairs: **200** (top)/Data from the Department of Home Affairs, drawn by Spatial Vision. • Food and Agriculture Organization of the United Nations: **272**/Source: Food and Agriculture Organization of the United Nations 2012 FAOSTAT, http://faostat3.fao.org/home//index.html • Fytogreen: **281** (bottom)/Fytogreen.com • Geography Teachers Association: **64** (bottom)/Geography Teachers Association of Victoria Inc. Interaction, journal of the GTAV, June 1998. Illustration redrawn by Harry Slaghekke. • Geoscience Australia: **29** (left), **29** (right)/Commonwealth of Australia Geoscience Australia 2012. This product is released under the Creative Commons Attribution 3.0 Australia Licence. • Getty Images: **34**/AFP; **59**/Rob Blakers; **63** (bottom right)/Portland Press Herald; **84** (bottom right)/Mlenny; **89** (middle right)/Auscape; **89** (bottom right)/Konrad Wothe/LOOK-foto; **98** (top left)/Universal Images Group; **111** (bottom)/guenterguni; **129**/Asahi Shimbun; **145**/Kay Dulay; **173** (bottom left)/RAUL ARBOLEDA/AFP; **173** (bottom right)/Wolfgang Kaehler/Contributor; **186** (bottom)/bloodstone; **188** (bottom)/Penny Tweedie; **214**/Keren Su; **215** (e)/MissHibiscus; **215** (b)/PATRICK BAZ/AFP; **218** (top)/Chris Mellor; **246** (bottom)/Paul Grand Image; **247** (top right)/sot; **276** (top)/Floris Leeuwenberg; **282** (bottom right)/pagadesign; **282** (middle right)/Steve Debenport; **283**/Nerthuz • Heritage Victoria: **56**/Heritage Victoria • International Association of Antarctica Tour Operators: **103**/International Association of Antarctica Tour Operators • iStockphoto: **282** (top left)/Rich Seymour • John Wiley & Sons Australia: **7** (top), **7** (bottom), **15**, **18**, **88**, **109**, **111** (top), **112** (top), **112** (bottom), **118**, **119**, **123**, **127** (top), **140**, **141** (top left), **141** (bottom left), **141** (top right), **141** (bottom right), **143**, **146**, **152** (bottom), **153**, **154**, **155**, **156** (top), **156** (bottom), **160** (top), **168** (top), **170** (top), **170** (middle), **172** (top), **172** (bottom),

199 (top), **205**, **212**, **234**, **261** (bottom right), **270** (bottom right); **51** (top)/Copyright © 2019 BBC; **255** (top)/Price Shire Council • Judy Mraz: **96** • Land Information NZ: **139**/Sourced from Topographic Map 273-09 Egmont. Crown Copyright Reserved. MAPgraphic Pty Ltd, Brisbane • Les ORourke: **279** (bottom)/ Les ORourke Photography • LSE Cities: **235** (top), **235** (bottom)/© LSE Cities/Urban Age Programme. "South American Cities: Securing an Urban Future" Urban Age Conference Newspaper, 2008, https:// LSECiti.es/u48af135 • MACRO ASSOCIATES: **203**/Original data of the graph from Department of Immigration • MAPgraphics: **5** (top), **6** (top), **29** (top), **31** (left), **80**, **137** (right), **151**, **157** (top), **158**, **163**, **164** (left), **165** (top left), **186** (top), **187** (bottom), **194**, **207**, **238**, **250**; **102**/© Geography Teachers Association of Victoria Inc • Mark Lincoln: **132**/Mark Lincoln • Michael Amendolia: **95**/© Michael Amendolia • NASA: **74**/JPL-Caltech / University of Colorado; **75** (right), **159** (bottom), **237** (top) • NASA Earth Observatory: **138** (right)/Image Science and Analysis Laboratory, NASA-Johnson Space Center; **225** (top)/Image by Craig Mayhew and Robert Simmon, NASA GSFC; **228** (top)/NASA; **232** (bottom left)/ NASA Earth Observatory • Newspix: **157** (bottom)/Anna Rogers; **269**/Glenn Daniels • OECD: **218** (bottom)/OECD Data, 2018, Foreign-Born Population, https://data.oecd.org/migration/foreign-born-population.htm. • PARS International: **274** (top)/From The New York Times, September 26, 2010 © 2010 The New York Times. All rights reserved. Used by permission and protected by the Copyright Laws of the United States. The printing, copying, redistribution, or retransmission of this Content without express written permission is prohibited • Paul F. Downton: **252** (bottom)/Perspective Sketch & Design by Paul F. Downton • Photodisc: **159** (top left), **159** (top right), **176** • RecycleWorks: **66**/Adapted from an image by RecycleWorks www.RecycleWorks.org • Regional Plan Association: **196**/Regional Plan Association • Science Photo Library: **98** (top left)/MDA Information Systems/Science Photo Library • Shutterstock: **1**/Paul Garnier Rimolo; **2**/wavebreakmedia; **3** (top left)/Toa55; **3** (top middle)/lightpoet; **3** (top right)/ bikeriderlondon; **3** (bottom left)/LDprod; **3** (bottom middle)/SpeedKingz; **3** (bottom right)/goodluz; **10** (bottom), **37**, **258** (bottom)/Neale Cousland; **11**, **259**/Nils Versemann; **14**/kojihirano; **16** (top)/Daniel Prudek; **16** (bottom), **48** (top)/Jason Patrick Ross; **19** (top)/Sarah Fields Photography; **25** (top)/Giancarlo Liguori; **35**/Xavier MARCHANT; **40**/Alexandra Martynova; **45**/John Le; **47**/tomas del amo; **48** (bottom)/ Alberto Loyo; **54**/worldroadtrip; **58**/Peter Clark; **60**/Preto Perola; **72**/Tupungato; **75** (left)/Alexey Fateev; **76** (bottom)/Tero Hakala; **78**/Sourav and Joyeeta; **79** (top)/Eniko Balogh; **79** (bottom)/linma; **83**/Roberto Caucino; **84** (top)/Bill Lawson; **84** (bottom left)/Anderl; **85** (top)/AustralianCamera; **85** (bottom)/Armin Rose; **89** (middle left)/Frantisek Staud; **89** (bottom left)/Ralph Griebenow; **90** (bottom)/Christopher Halloran; **92** (bottom)/N Mrtgh; **101** (bottom)/steve estvanik; **105**/Makhnach_S; **107**/Dominik Michalek; **113** (top)/John A Cameron; **115**/Francesco Carucci; **113** (bottom left)/Denis Mironov; **113** (bottom right)/ Josemaria Toscano; **120** (top)/Craig Hanson; **120** (bottom)/CO Leong; **126**/NigelSpiers; **133** (top)/ jakubtravelphoto; **137** (left)/Bjartur Snorrason; **138** (left)/bartuchna@yahoo.pl; **152** (top left)/ AustralianCamera; **152** (top middle)/Ashley Whitworth; **152** (top right)/Paulo Vilela; **160** (bottom)/Dirk Ercken; **161** (right)/Ammit Jack; **161** (left)/Frontpage; **161** (right bottom inset)/Lukas Gojda; **161** (right top inset)/OlegDoroshin; **167** (bottom)/YIFANG NIE; **168** (top)/Dr. Morley Read; **170** (bottom), **282** (top right)/Janelle Lugge; **175** (top)/Txanbelin; **178**/Sergey Uryadnikov; **181**/littlesam; **183** (right)/John-james Gerber; **208**/Edward Haylan; **209** (bottom)/Hung Chung Chih; **210**/BartlomiejMagierowski; **215** (a)/Sadik Gulec; **215** (c)/Earl D. Walker; **215** (d)/Shukaylova Zinaida; **215** (f)/forestpath; **215** (g)/africa924; **216** (a)/Daxiao Productions; **216** (b)/Asianet-Pakistan; **216** (c)/BartlomiejMagierowski; **216** (d)/Jane September; **216** (e)/Elzbieta Sekowska; **216** (g)/De Visu; **219** (bottom)/Authentic travel; **222** (top)/egd; **224**/Tomasz Szymanski; **226**/Ildi Papp; **228** (bottom)/fuyu liu; **232** (top)/amadeustx; **233** (top)/Kzenon; **236** (top)/Andrea Dal Max; **240** (top)/Andrew Zarivny; **242** (top right)/Rostislav Glinsky; **242** (bottom)/ slava17; **243** (left)/Keith Wheatley; **243** (right)/Bikeworldtravel; **247** (bottom)/lucarista; **255** (bottom)/ Andrew Zarivny; **257**/stocker1970; **268** (bottom)/CTR Photos; **270** (top)/paintings; **276** (bottom)/ Aleksandar Todorovic; **281** (top)/Sunflowerey; **282** (middle left)/THPStock; **282** (bottom left)/Blazej Lyjak • Spatial Vision: **28**, **31** (right), **33**, **38**, **57** (top), **57** (bottom left), **69**, **110**, **117**, **127** (bottom), **128** (top), **130**, **136** (top), **138**(top), **171** (top), **171** (bottom), **189** (bottom), **190**, **197**, **230**, **231**, **239** (bottom), **241**, **263**,

275 (top); **6** (bottom), **189**/Data source from ABS. Map by Spatial Vision; **8**, **265**/Data source from Department of Environment, Land, Water and Planning © State Government of Victoria 2016. Map by Spatial Vision; **10** (top), **258** (top)/© OpenStreetMap contributors and Spatial Vision; **25** (bottom)/Source: World Map of Carbonate Rock Outcrops v3.0. Map created by Spatial Vision.; **42**/Copyright © 1992–2012 UNESCO/World Heritage Centre. All rights reserved. Map created by Spatial Vision.; **42** (top), **91**, **95**, **108**, **125** (bottom), **188** (top), **199** (bottom), **209** (top), **237** (bottom), **273**, **279**/© Spatial Vision; **182** (bottom)/World Bank Data; **191**/Sources: Various Victorian planning studies and current land use mapping. Spatial Vision; **202**/Sydney Morning Herald, drawn by Spatial Vision; **222** (bottom)/Data from Our World in Data, drawn by Spatial Vision.; **262** (top)/Created using data from World Bank • Sustainable Cities Internat.: **246** (top)/Photo by Sumana Wijeratne, VanLanka Planning • Terry McMeekin: **242** (top left)/Terry McMeekin • U.S. Geological Survey: **125** (top left)/U.S. Geological Survey • UN-Habitat United Nations Human Settlements Programme: **219** (top)/Derived from UN-HABITAT The State of African Cities 2010 • United Nations: **182** (top)/From World Population Prospects The 2015 Revision, Key Findings and Advance Tables by DESA, Population Division, c 2015 United Nations. Reprinted with the permission of the United Nations.; **225** (bottom)/© 2018 United Nations, DESA, Population Division. Licensed under Creative Commons license CC BY 3.0 IGO.; **227**/United Nations, Department of Economic and Social Affairs, Population Division 2018. The World's Cities in 2018—Data Booklet ST/ESA/ SER.A/417. • USGS National Center: **114**, **117**, **125** (top right), **136** (bottom)/US Geological Survey • Wikimedia Commons: **70**/US Army Corp of Engineers.

Text

• Australian Bureau of Statistics: **189** (top), **200** (bottom), **201** (top), **201** (bottom) • Benjamin Baird: **122**/Benjamin Baird • Creative Commons: **190**/© Infrastructure Australia 2015; **203**/Commonwealth of Australia; **204**/Australian Government, Department of Immigration and Border Protection; **221** (left)/ Progress on drinking water, sanitation and hygiene: 2017 update and SDG baselines. Geneva: World Health Organization WHO and the United Nations Children's Fund UNICEF, 2017. Licence: CC BY-NC-SA 3.0 IGO; **225** (right)/Progress on drinking water, sanitation and hygiene: 2017 update and SDG baselines. Geneva: World Health Organization WHO and the United Nations Children's Fund UNICEF, 2017. Licence: CC BY-NC-SA 3.0 IGO • FSC International Center GmbH: **176** (bottom)/FSC • John Wiley & Sons Australia: **144**, **195**/Based on data from the ABS and US Population Bureau • Spatial Vision: **261**/United Nations and Spatial Vision • U.S. Census Bureau: **240** (bottom)/U.S. Census Bureau • United Nations: **221** • United Nations Department of Economic and Social Affairs: **187**/Data obtained from United Nations, Department of Economic and Social Affairs, Population Division 2015. World Population Prospects: The 2015 Revision.

Every effort has been made to trace the ownership of copyright material. Information that will enable the publisher to rectify any error or omission in subsequent reprints will be welcome. In such cases, please contact the Permissions Section of John Wiley & Sons Australia, Ltd.

1 Geographical skills and concepts

1.1 Overview

1.1.1 Introduction

As a student of Geography, you are starting to build the knowledge and skills that will be needed by you and your community now and into the future. The concepts and skills that you will use will not only help you in Geography but they can also be applied to everyday situations, such as finding your way from one place to another. Studying Geography may even help you in a future career here in Australia or somewhere overseas.

Throughout your study of Geography you will cover topics that will give you a better understanding of the world around you — both the local and global environment. You will investigate issues that need to be addressed now and also options for the future.

 Resources

eWorkbook Customisable worksheets for this topic

LEARNING SEQUENCE

1.1 Overview
1.2 Work and careers in Geography
1.3 Concepts and skills used in Geography

1.4 Review   online only

To access interactivities and resources, select your learnON format at www.jacplus.com.au.

1.2 Work and careers in Geography

1.2.1 Links to Geography

Many questions come up during a typical Geography class, such as the ones in **TABLE 1**. These questions need to be answered in the real world by people in a wide variety of occupations. They all have links with Geography.

TABLE 1 Examples of occupations that use Geography

Question	Occupations/organisations that try to answer these questions
How high *is* Mount Everest? How do we know?	Surveyor, Cartographer
How can we protect our parks and wildlife?	Park ranger, Planner, Environmental manager
Where should we establish a new suburb for our future population?	Urban planner, Demographer
How can we prepare for future droughts and floods?	Civil engineer
Does our town really have enough water? Should we build a new dam? Where should we build a new dam?	Coastal engineer, Hydrologist, Cartographer
Should a boat marina be built at location X or at location Y?	Oceanographer
Do we have good quality drinking water?	Chemist, Hydrologist
How do countries such as India and China deal with their air pollution problems?	Environmental scientist/manager
How do we provide aid to other countries?	Air Force, Navy, Army Officer. Red Cross, World Vision and other aid agencies.
How do we build sustainable housing?	Architect, Landscape architect, Civil engineer/Construction manager, Town planner, Real estate salesperson

Think: who are you and what is your position in the world?

Do you know much about the occupations mentioned in **TABLE 1**? Are any of interest to you?

The first step in thinking about your future is to consider questions such as:

- Who am I?
- What are my interests?
- What do I enjoy doing?
- What am I good at?
- What would I like to do when I leave school?

1.2.2 Geography careers on the move

A great part of studying Geography is being able to explore the many occupations and areas that it opens up. In **TABLE 2** are some occupations that you may not have thought studying Geography could lead you into.

TABLE 2 Would I enjoy . . .

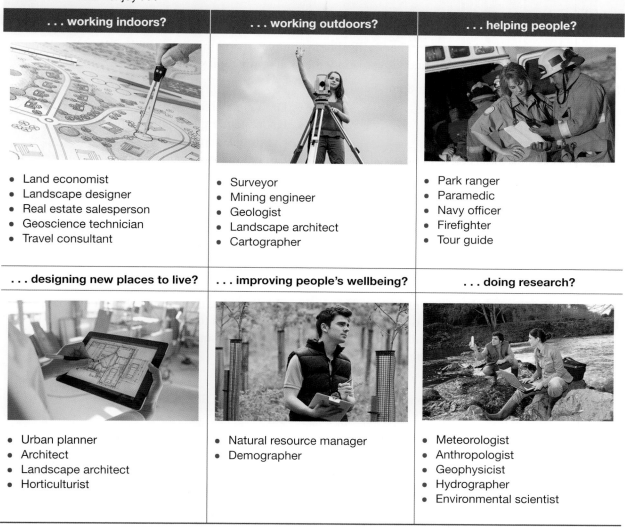

. . . working indoors?	. . . working outdoors?	. . . helping people?
• Land economist • Landscape designer • Real estate salesperson • Geoscience technician • Travel consultant	• Surveyor • Mining engineer • Geologist • Landscape architect • Cartographer	• Park ranger • Paramedic • Navy officer • Firefighter • Tour guide
. . . designing new places to live?	**. . . improving people's wellbeing?**	**. . . doing research?**
• Urban planner • Architect • Landscape architect • Horticulturist	• Natural resource manager • Demographer	• Meteorologist • Anthropologist • Geophysicist • Hydrographer • Environmental scientist

1.2.3 Finding my way as a local and global citizen

A wide range of exciting new jobs are developing in the spatial sciences which use geographical tools such as GPS, GIS, satellite imaging and surveying. These tools help people make important decisions about managing and planning places and resources. Whether it be how to manage water somewhere in the Middle East or how best to design a new housing estate here in Australia, these skills and occupations will be an important part of working as a global citizen.

1.3 Concepts and skills used in Geography

1.3.1 Skills used in studying Geography

As you work through each of the topics in this title, you'll complete a range of exercises to check and apply your understanding of concepts covered. In each of these exercises, you'll use a variety of skills, which are identified using the Geographical skills (GS) key provided at the start of each exercise set.

The skills are:
- **GS1** Remembering and understanding
- **GS2** Describing and explaining
- **GS3** Comparing and contrasting
- **GS4** Classifying, organising, constructing
- **GS5** Examining, analysing, interpreting
- **GS6** Evaluating, predicting, proposing

In addition to these broad skills, there is a range of essential practical skills that you will learn, practise and master as you study Geography. The SkillBuilder subtopics found throughout this title will tell you about the skill, show you the skill and let you apply the skill to the topics covered.

The SkillBuilders you'll use in Year 8 are listed below.

- Recognising land features
- Using positional language
- Constructing a field sketch
- Reading contour lines on a map
- Using latitude and longitude
- Calculating distance using scale
- Drawing simple cross-sections
- Interpreting an aerial photo
- Creating and describing complex overlay maps

- Drawing a précis map
- Understanding thematic maps
- Creating and reading pictographs
- Comparing population profiles
- Describing photographs
- Creating and reading compound bar graphs
- Constructing a basic sketch map
- Reading and describing basic choropleth maps
- Drawing a line graph using Excel

1.3.2 SPICESS

Geographical concepts help you to make sense of your world. By using these concepts you can both investigate and understand the world you live in, and you can use them to try to imagine a different world. The concepts help you to think geographically. There are seven major concepts: ***space, place, interconnection, change, environment, sustainability*** and ***scale***.

FIGURE 1 A way to remember these seven concepts is to think of the term SPICESS.

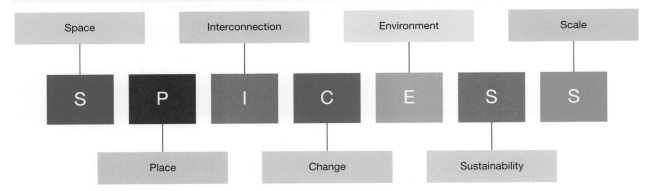

You will use the seven concepts to investigate two units: *Landforms and landscapes* and *Changing nations*.

1.3.3 What is space?

Everything has a location on the space that is the surface of the Earth, and studying the effects of location, the distribution of things across this space, and how the space is organised and managed by people, helps us to understand why the world is like it is.

A place can be described by its absolute location (latitude and longitude) or its relative location (in what direction and how far it is from another place).

FIGURE 2 The distribution of the world's deserts

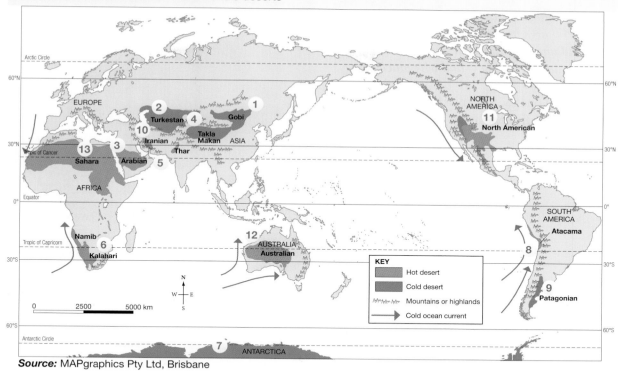

Source: MAPgraphics Pty Ltd, Brisbane

Explore more with my**World**Atlas

Deepen your understanding of this topic with related case studies and questions.
• Developing Australian Curriculum concepts > **Space**

1.3.4 What is place?

The world is made up of places, so to understand our world we need to understand its places by studying their variety, how they influence our lives and how we create and change them.

You often have mental images and perceptions of places — rich and poor cities, suburbs, towns or neighbourhoods — and these may be very different from someone else's perceptions of the same places.

FIGURE 3 The Paraisópolis favela (slum), home to 60 000 people, is situated next to the gated complexes of the wealthy Morumbi district of São Paulo.

Explore more with my**World**Atlas

Deepen your understanding of this topic with related case studies and questions.
• Developing Australian Curriculum concepts > **Place**

1.3.5 What is interconnection?

People and things are connected to other people and things in their own and other places, and understanding these connections helps us to understand how and why places are changing.

An event in one location can lead to change in a place some distance away.

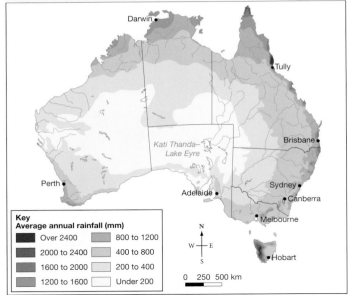

FIGURE 4 Distribution of annual rainfall in Australia

Source: MAPgraphics Pty Ltd, Brisbane

FIGURE 5 Australia's population distribution and density

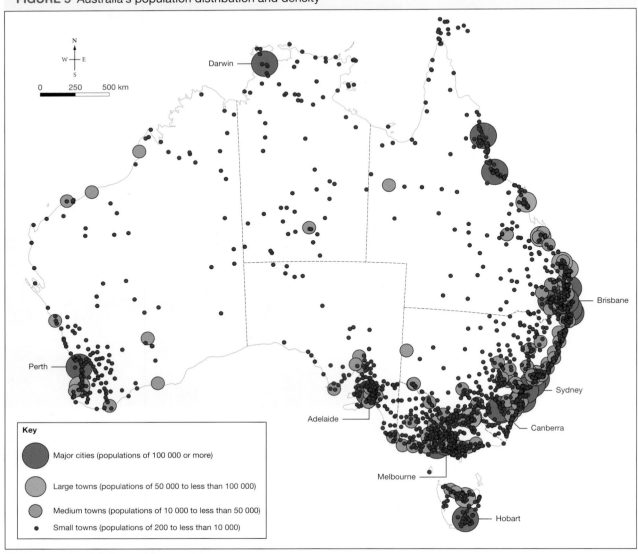

Source: Australian Bureau of Statistics

─Explore more with my**World**Atlas──────────────

Deepen your understanding of this topic with related case studies and questions.
- Developing Australian Curriculum concepts > **Interconnection**

1.3.6 What is change?

The concept of change is about using time to better understand a place, an environment, a spatial pattern or a geographical problem.

Some changes can be fast and easily observed, but others are very slow. Cities, for example, can expand outwards over a number of years. Similarly, landforms generally change very slowly, as with the formation of mountains. But some landscape change can be very fast, as is the case with landslides, volcanic eruptions and deforestation.

FIGURE 6a Landscape before deforestation

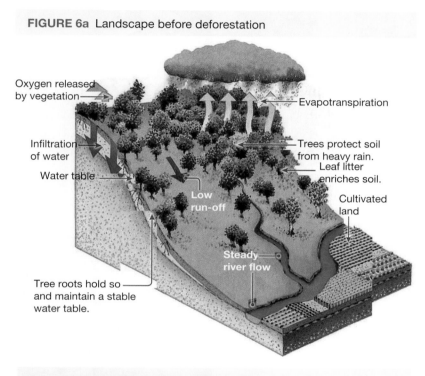

Oxygen released by vegetation

Evapotranspiration

Infiltration of water

Trees protect soil from heavy rain.
Leaf litter enriches soil.

Water table

Low run-off

Cultivated land

Steady river flow

Tree roots hold so and maintain a stable water table.

FIGURE 6b Landscape after deforestation

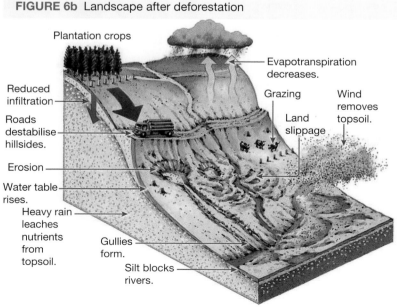

Plantation crops

Evapotranspiration decreases.

Reduced infiltration

Grazing

Wind removes topsoil.

Roads destabilise hillsides.

Land slippage

Erosion

Water table rises.

Heavy rain leaches nutrients from topsoil.

Gullies form.

Silt blocks rivers.

FIGURE 7 The history of Melbourne's urban sprawl

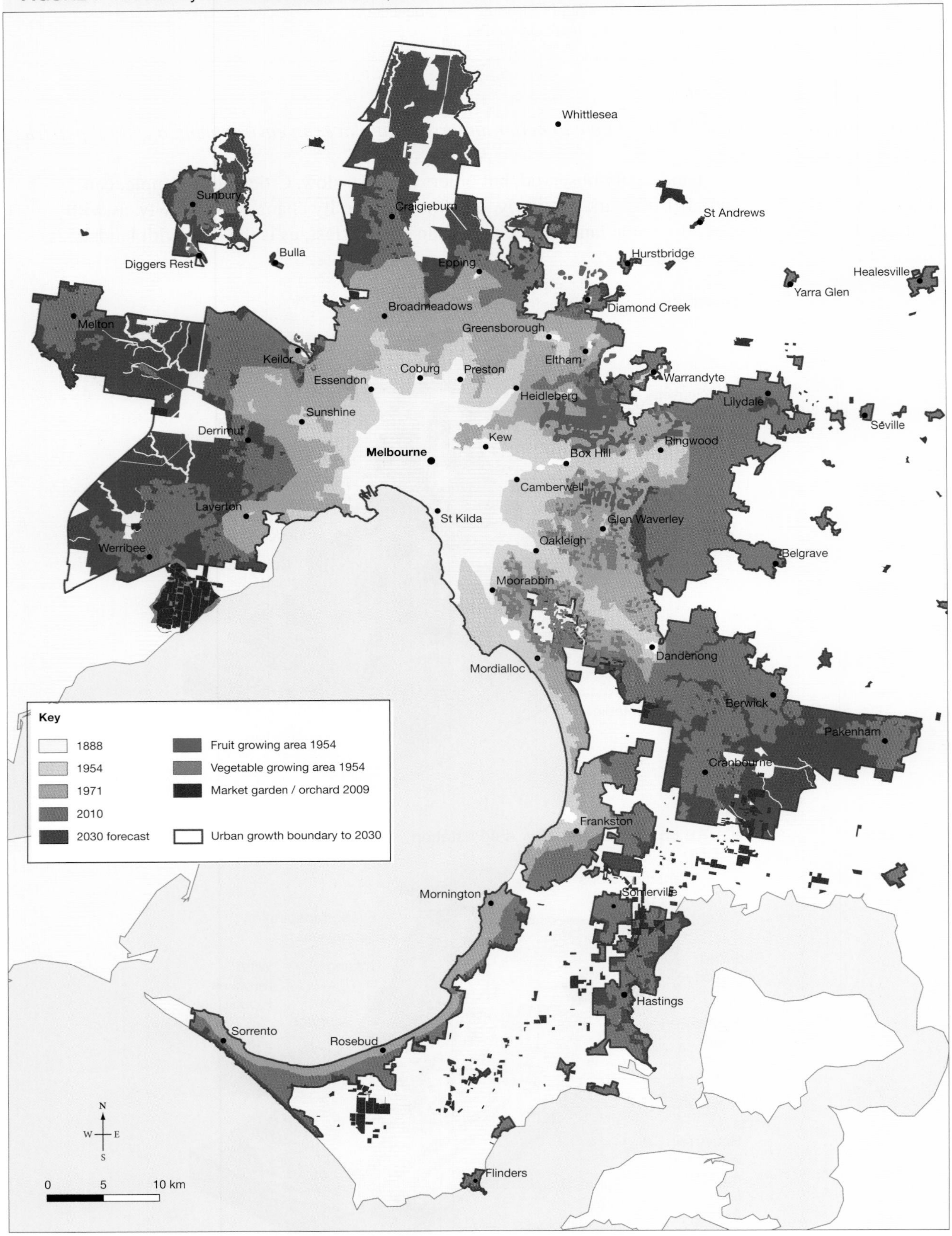

Key

1888	Fruit growing area 1954
1954	Vegetable growing area 1954
1971	Market garden / orchard 2009
2010	Urban growth boundary to 2030
2030 forecast	

Source: Various Victorian planning studies and current land use mapping. Map produced by Spatial Vision 2019.

┌─Explore more with my**World**Atlas───

Deepen your understanding of this topic with related case studies and questions.
• Developing Australian Curriculum concepts > **Change**

1.3.7 What is environment?

People live in and depend on the environment, so it has an important influence on our lives.

The environment, defined as the physical and biological world around us, supports and enriches human and other life by providing raw materials and food, absorbing and recycling wastes, and being a source of enjoyment and inspiration to people.

1.3.8 What is sustainability?

Sustainability is about maintaining the capacity of the environment to support our lives and those of other living creatures.

Sustainability is about the interconnection between the human and natural world and who gets which resources and where, in relation to conservation of these resources and prevention of environmental damage.

FIGURE 8 Uranium mining in Colorado, United States. Many deserts contain valuable mineral deposits.

FIGURE 9 The Vatican is the world's smallest independent state. In 2008 more than 2000 photovoltaic panels were fixed to the roof of one of the city state's main buildings — the roof of the Paul VI Hall — enabling the Vatican to cut its carbon dioxide emissions by about 225 tonnes a year. The 2400 panels heat, light and cool the hall and several surrounding buildings, producing 300 kilowatt hours (MWh) of clean energy a year.

┌─Explore more with my**World**Atlas───

Deepen your understanding of this topic with related case studies and questions.
• Developing Australian Curriculum concepts > **Environment**
• Developing Australian Curriculum concepts > **Sustainability**

1.3.9 What is scale?

When we examine geographical questions at different spatial levels we are using the concept of scale to find more complete answers.

Scale can be from personal and local to regional, national or global. Looking at things at a range of scales allows a deeper understanding of geographical issues.

Ways to improve sustainability at the local scale include:

- reducing the ecological footprint
- protecting the natural environment
- increasing community wellbeing and pride in the local area
- changing behaviour patterns by providing better local options
- encouraging compact or dense living
- providing easy access to work, play and schools.

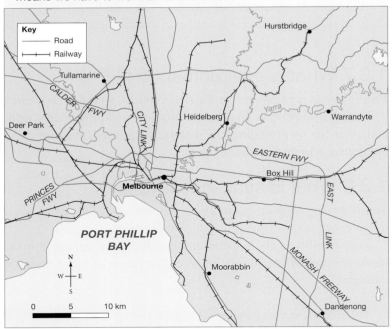

FIGURE 10 Melbourne, Victoria. Building sustainable communities means we have to work at various scales.

Source: Department of Environment, Land, Water and Planning

FIGURE 11 Street art in Melbourne, Australia

Ways to improve sustainability at the city scale include:

- building strong central activities areas (either one major hub, or a number of specified activity areas)
- reducing traffic congestion
- protecting natural systems
- avoiding suburban sprawl and reducing inefficient land use
- distributing infrastructure and transport networks equally and efficiently to provide accessible, cheap transportation options
- promoting inclusive planning and urban design
- providing better access to healthy lifestyles (e.g. cycle and walking paths)
- improving air quality and waste management
- using stormwater more efficiently
- increasing access to parks and green spaces
- reducing car dependency and increasing walkability
- promoting green space and recreational areas
- demonstrating a high mix of uses (e.g. commercial, residential and recreational).

FIGURE 12 The Melbourne skyline with the Melbourne Sports and Entertainment Precinct in the foreground

┌─ Explore more with my**World**Atlas ──────────────────────────

Deepen your understanding of this topic with related case studies and questions.
- Developing Australian Curriculum concepts > **Scale**

1.3 INQUIRY ACTIVITIES

1. Refer to **FIGURE 2** and an atlas to answer the following.
    (a) Give the absolute location (latitude and longitude) of Mecca, in the Arabian desert. What is the relative location of Mecca from Australia?
    (b) Describe the *spatial* distribution of the world's deserts in relation to the tropics.
    (c) In what direction and approximately how far is the Thar Desert from the Arabian Desert, the Atacama Desert and the Namib Desert?
    (d) How is the location of the Namib Desert influenced by cold ocean currents?

    Examining, analysing, interpreting

2. Describe any local action where you live that tries to improve *sustainability*. You could talk to your parents about this, or contact your local council to see what they are trying to do about the issue.

    Describing and explaining

1.4 Review

1.4.1 Key knowledge summary

Use this dot point summary to review the content covered in this topic.

1.4 Exercise 1: Review

Select your learnON format to complete review questions for this topic.

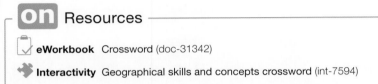

Resources

  eWorkbook Crossword (doc-31342)

  Interactivity Geographical skills and concepts crossword (int-7594)

UNIT 1
LANDFORMS AND LANDSCAPES

Have you ever stood on a hill, or high ground, and looked at the scenery and landscape in front of you? From a height you can see a variety of different landforms such as mountains, valleys and plains. So, how are different landforms actually created? And what causes the hazards we need to deal with?

FIELDWORK INQUIRY: LOCAL WATER CATCHMENT STUDY

Your task

Your team has been commissioned by the local water authority to compile and present a report evaluating the current state of your local catchment. Your team must gather data to investigate how the catchment changes from the upper reaches to the lower.

Select your learnON format to access:

- an overview of the project task
- details of the inquiry process
- resources to guide your inquiry
- an assessment rubric.

on Resources

ProjectsPLUS Fieldwork inquiry: Local water catchment study (pro-0145)

2 Introducing landforms and landscapes

2.1 Overview

From oceans to deserts to cities, what exactly are landscapes and how is each one unique?

2.1.1 Introduction

World landscapes and landforms

Landscapes are the visible features of the land, ranging from the icy landscapes of polar regions and lofty mountain ranges, through to forests, deserts and coastal plains. Shaped by physical processes over millions of years, they have been overlaid by the presence of humans; this includes the places we build, such as towns and cities, and the changes we make to the natural landscape.

 Resources

✓ **eWorkbook**	Customisable worksheets for this topic	
🎞 **Video eLesson**	World landscapes and landforms (eles-1623)	

LEARNING SEQUENCE

- **2.1** Overview
- **2.2** Different types of landscapes
- **2.3** **SkillBuilder:** Recognising land features ⬛online only
- **2.4** The processes that shape landscapes
- **2.5** Underground landscapes
- **2.6** Australian landforms
- **2.7** Landforms of the Pacific
- **2.8** **SkillBuilder:** Using positional language ⬛online only
- **2.9** Cultural significance of landscapes
- **2.10** Preserving and managing landscapes
- **2.11** **Thinking Big research project**: Karst landscape virtual tour ⬛online only
- **2.12** **Review** ⬛online only

To access a pre-test and starter questions and receive immediate, **corrective feedback** and **sample responses** to every question, select your learnON format at www.jacplus.com.au.

2.2 Different types of landscapes

2.2.1 Types of landscapes

There are many different landscapes across the Earth, and similarities can be observed within regions. Variations in landscapes are influenced by factors such as: climate; geographical features, including mountains and rivers; latitude; the impact of humans; and where the landscapes are located.

FIGURE 1 Selected world landscapes

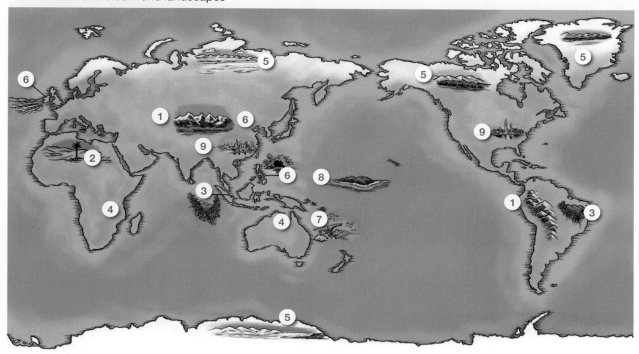

1. Mountains rise above the surrounding landscape. They often have steep sides and high peaks and are the result of processes operating deep inside the Earth. Some reach high into the atmosphere where it is so cold that snow is found on their peaks.

2. Deserts are areas of low rainfall; they are an arid or dry environment. They can experience temperature extremes: hot by day and freezing at night. However, not all deserts are hot. Antarctica is the world's largest desert, and the Gobi Desert, located on a high **plateau** in Asia, is also a cold desert.

3. Rainforests are the most diverse landscapes on Earth. They are found in a variety of climates, ranging from the hot wet tropics to the cooler temperate areas. The lush vegetation found in these regions depends on a high level of rainfall. Over 50 per cent of all known plant and animal species are found within them. In addition, many of our foods and medicines come from rainforests.

4. Grasslands, or savannas, are sometimes seen as a transitional landscape found between forests and deserts. They contain grasses of varying heights and coarseness, and small or widely spaced trees. They are often inhabited by grazing animals.

5. Polar regions and tundra can be found in polar and alpine regions. Characterised by **permafrost**, they are too cold for trees to grow. Vegetation such as dwarf shrubs, grasses and lichens have adapted to the extreme cold and short growing season. **Glaciers** often carve spectacular landscape features.

6. Karst landscapes form when mildly acidic water flows over soluble rock such as limestone. Small fractures form, which increase in size over time and lead to underground drainage systems developing. Common landforms include limestone pavements, disappearing rivers, reappearing springs, sinkholes, caves and karst mountains. Around 25 per cent of the world's population obtains water from karst **aquifers**.

7. Aquatic landscapes cover around three-quarters of the Earth and can be classified as freshwater or marine. Marine landscapes are the saltwater regions of the world, and include oceans and coral reefs. Freshwater landscapes are found on land, and include lakes, rivers and wetlands.

8. Islands are areas of land that are completely surrounded by water. They can be continental or oceanic. Continental islands lie on a continental shelf — an extension of a continent that is submerged beneath the sea. Oceanic islands rise from the ocean floor and are generally volcanic in origin. A group or chain of islands is known as an archipelago.

9. Human or built landscapes are those that have been altered or created by humans.

FIGURE 2 At 8848 metres, Mount Everest in the Himalayas is the highest mountain on Earth.

FIGURE 3 These cave formations in Alabama are protected.

on Resources

Interactivity   Landscapes galore (int-3102)

Google Earth   Mount Everest

Explore more with my**World**Atlas

Deepen your understanding of this topic with related case studies and questions.
- Investigating Australian Curriculum topics > Year 8: Landforms and landscapes > **Grasslands**

2.2 INQUIRY ACTIVITY

Investigate one of the landscapes featured in **FIGURE 1** and find out some *places* in which it is found. Show this information on a map. Annotate your map with information from this subtopic and characteristics of your landscape.   **Classifying, organising, constructing**

2.2 EXERCISES

Geographical skills key: GS1 Remembering and understanding **GS2** Describing and explaining **GS3** Comparing and contrasting **GS4** Classifying, organising, constructing **GS5** Examining, analysing, interpreting **GS6** Evaluating, predicting, proposing

2.2 Exercise 1: Check your understanding

1. **GS2** Describe a built *environment*. What do you think a natural *environment* is?
2. **GS1** What factors make landscapes different?
3. **GS1** List as many different human or built *environments* as you can think of.
4. **GS2** Why do you think people *change* landscapes?
5. **GS2** Select two of the landscapes featured in this subtopic and explain how they are different.

2.2 Exercise 2: Apply your understanding

1. **GS4** Copy the following table into your workbook.

Characteristics	How people use it	Positive impacts	Negative impacts

    (a) Select one of the landscape types described in this subtopic and complete the table, noting the positive and negative aspects of human use.
    (b) Which list is larger — the positive impacts or negative impacts?
    (c) Review the column of negative impacts. Select three of these impacts and suggest a way in which the *environment* could be used more *sustainably*.

2. **GS2** Describe how the *scale* of the following landscapes might differ around the world: deserts, polar regions, aquatic landscapes and islands.
3. **GS5** Which of the featured landscapes would you like to know more about? Draw up a list of questions that you would like to have answered.
4. **GS5** Why do you think rainforests are described as 'the most diverse landscapes on Earth'?
5. **GS5** Which of the featured landscapes do you think would be the least diverse? Give reasons for your answer.

Try these questions in learnON for instant, corrective feedback. Go to www.jacplus.com.au.

2.3 SkillBuilder: Recognising land features

What are land features?

Land features are landforms with distinct shapes, such as hills, valleys and mountains. You can recognise these as you look around your natural environment. On topographic maps you can recognise land features from the patterns formed by the contour lines.

Select your learnON format to access:

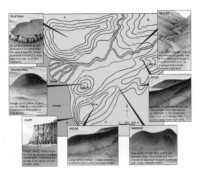

- an overview of the skill and its application in Geography (Tell me)
- a video and a step-by-step process to explain the skill (Show me)
- an activity and interactivity for you to practise the skill (Let me do it)
- questions to consolidate your understanding of the skill.

---

on Resources

Video eLesson SkillBuilder: Recognising land features (eles-1648)

Interactivity SkillBuilder: Recognising land features (int-3144)

2.4 The processes that shape landscapes

2.4.1 Are all processes natural?

There are processes at work that continuously sculpt and change the landscape. In the future, the Earth's surface will look very different from the way it looks today.

There are a variety of natural processes that shape and reshape not only the surface of the Earth, but also what lies beneath it. Natural processes include uplift, such as that caused by tectonic activity, **erosion**, **deposition** and **weathering**. People change the landscape when they clear land for agriculture or build cities and road networks. Sometimes they alter the course of a river or trap its flow behind the walls of a dam.

2.4.2 The role of tectonic forces

The Earth's surface, or crust, is split into a number of plates, which fit together like a giant jigsaw puzzle. These plates sit on a layer of semi-molten material in the Earth's **mantle** — the layer of the Earth between the crust and the core. Heat from the Earth's core creates convection currents within the mantle, causing the plates to move. Most of the Earth's great mountain regions were formed as a result of this movement.

When two plates collide, one plate often slides under the other, in a process known as subduction, and it becomes part of the mantle. Other rocks are forced upwards and bent or folded. Large mountain ranges that were formed in this way include the Himalayas in Asia and the Rocky Mountains in North America. You will find more information on how mountains are formed in subtopic 5.2.

2.4.3 How is the landscape worn away?

Erosion is the wearing away of the Earth's surface by natural elements such as wind, water, ice and human activity. The landscape is further eroded when agents such as wind, water and ice **transport** these materials to new locations. Eventually, transported material is deposited in a new location. Over time, this material can build up and new landforms result. The Grand Canyon in Colorado in the United States (**FIGURE 1**) is an example of these elements at work. These processes work more quickly on softer rocks.

Human activity also contributes to erosion. Deforestation, agriculture, urban sprawl, logging and road construction all alter the natural balance and increase erosion by as much as 40 per cent in some areas. Vegetation not only provides valuable habitat for native animals but is also vital for binding the soil together. Once vegetation is removed, it is more easily broken down and removed by wind and water. When topsoil (see **FIGURE 6**) is removed, plants are unable to obtain the nutrients they need for growth. Sometimes wide, deep channels, known as gullies, form (**FIGURE 2**).

FIGURE 1 Over millions of years, the Colorado River has cut deep channels to form the Grand Canyon.

FIGURE 2 Note the scale of this gully compared to the people.

FIGURE 3 After tectonic forces cause a section of the Earth to be raised (uplifted), other processes take over and resculpt the landscape.

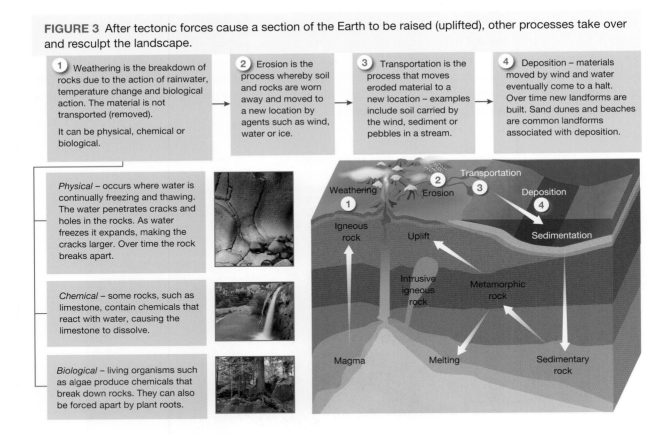

1 Weathering is the breakdown of rocks due to the action of rainwater, temperature change and biological action. The material is not transported (removed).

It can be physical, chemical or biological.

2 Erosion is the process whereby soil and rocks are worn away and moved to a new location by agents such as wind, water or ice.

3 Transportation is the process that moves eroded material to a new location – examples include soil carried by the wind, sediment or pebbles in a stream.

4 Deposition – materials moved by wind and water eventually come to a halt. Over time new landforms are built. Sand dunes and beaches are common landforms associated with deposition.

Physical – occurs where water is continually freezing and thawing. The water penetrates cracks and holes in the rocks. As water freezes it expands, making the cracks larger. Over time the rock breaks apart.

Chemical – some rocks, such as limestone, contain chemicals that react with water, causing the limestone to dissolve.

Biological – living organisms such as algae produce chemicals that break down rocks. They can also be forced apart by plant roots.

on Resources

Interactivity   Break down! (int-3101)
Google Earth   Grand Canyon

Explore more with my**World**Atlas

Deepen your understanding of this topic with related case studies and questions.
- Investigate additional topics > Earthquakes and volcanoes > **Active Earth**

2.4.4 What is soil?

We rarely give much thought to the soil beneath our feet. But soil is the basis of all life on the Earth. It provides the nutrients needed for growing plants, which provide food for animals. Without soil, people could not grow crops or raise livestock. Without soil, nothing could survive.

Soil is a thin layer of material on the surface of the Earth. In it, plants can grow. In some parts of the world it is metres deep, but in Australia it is a thin layer of 15 to 20 centimetres depth. The composition of soil is shown in **FIGURE 4** and the factors that influence soil formation are shown in **FIGURE 5**.

Australia generally has poor soils when compared with those found on other continents such as North America and Europe. Australian soils are generally low in nutrients and, in some areas, especially arid zones, they have a high salt content. Patches of good soil, though, are scattered throughout the continent.

For example, there is:
- volcanic soil on the Darling Downs in Queensland and around Orange in New South Wales
- alluvial soil in river valleys such as around the Clarence River in New South Wales and Margaret River in Western Australia.

In many parts of Australia, it takes more than 1000 years for natural processes to produce three centimetres of soil.

How is soil formed?
Factors that influence soil formation are shown in **FIGURE 5**.

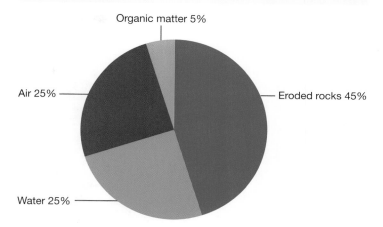

FIGURE 4 While the composition of soil varies widely across the Earth, an average soil will have these characteristics.

- Organic matter 5%
- Eroded rocks 45%
- Water 25%
- Air 25%

FIGURE 5 Influences on soil formation

Climate affects the rate of weathering of soil. In high rainfall areas, soil develops more rapidly, but excess moisture also washes out or leaches nutrients. In rainforests, for example, the rich supply of humus from decaying plant matter produces lush vegetation. However, high rainfall means that without this constant supply of humus, soil fertility is quickly lost. In arid regions, where evaporation is high, soils often contain too much salt to support plant growth. Weather also plays a role; a climate with a freezing and thawing cycle will speed up the breakdown of rocks. In warm climates, the activity of soil organisms is high, and chemical processes also happen more quickly.

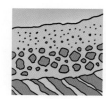

Surface rocks and bedrock are broken down through weathering and erosion. The type of soil that forms depends on the parent material and the minerals it contains. A coarse, sandy soil will develop from sandstone. Bedrock that is mainly granite produces a sandy loam, while shale turns into heavy clay soil.

Soil

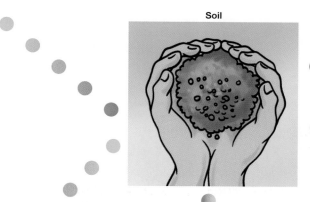

Plants and animals

Decaying plant and animal matter on the soil's surface is broken down by microorganisms into material that is incorporated into the soil, making it nutrient rich.

Topography

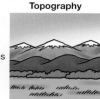

Surface features such as hills, valleys and rivers influence soil development. Soil is generally deeper on the top and at the base of a hill than on its slopes. Floodplains next to river valleys are often nutrient rich, due to sediment being deposited as floodwaters recede.

Time

These processes take place over long periods of time. Soils undergo many changes with the passage of time.

2.4.5 What is a soil profile?

Soil forms in layers called horizons (see **FIGURE 6**). A soil profile is a side view or cross-section of these different layers or horizons.

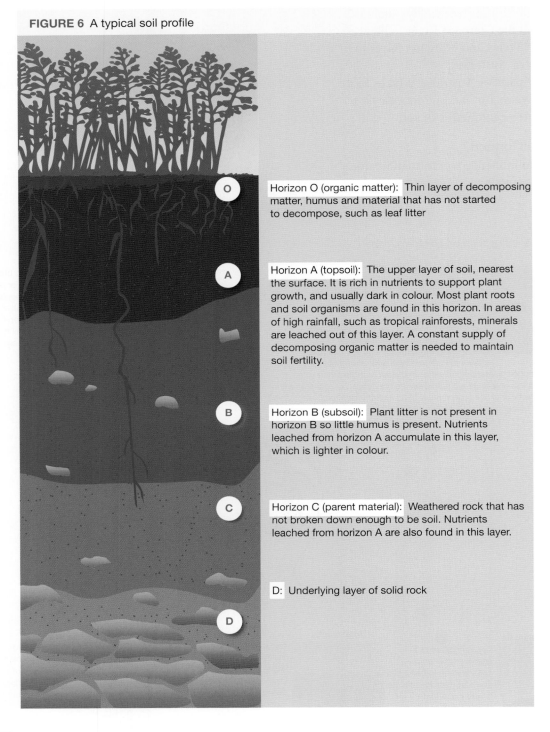

FIGURE 6 A typical soil profile

Horizon O (organic matter): Thin layer of decomposing matter, humus and material that has not started to decompose, such as leaf litter

Horizon A (topsoil): The upper layer of soil, nearest the surface. It is rich in nutrients to support plant growth, and usually dark in colour. Most plant roots and soil organisms are found in this horizon. In areas of high rainfall, such as tropical rainforests, minerals are leached out of this layer. A constant supply of decomposing organic matter is needed to maintain soil fertility.

Horizon B (subsoil): Plant litter is not present in horizon B so little humus is present. Nutrients leached from horizon A accumulate in this layer, which is lighter in colour.

Horizon C (parent material): Weathered rock that has not broken down enough to be soil. Nutrients leached from horizon A are also found in this layer.

D: Underlying layer of solid rock

 Resources

eWorkbook Soil formation

Weblink Soil formation

2.4 INQUIRY ACTIVITIES

1. Use the internet to discover a landform that you find interesting. Copy and paste an image of the landform into a Word document. Annotate your image with information about its location and how it was formed. Add further annotations to describe how your landform might have **changed** over time. Detail how it might have looked in the past and how it might look in the future. Think carefully about the **scale** of this **change**.

   Classifying, organising, constructing

2. Study the **environment** around your home or school and find a **place** where there is evidence of erosion. Make a sketch and label the features of the landscape. Highlight areas where erosion is evident and add annotations to explain what you think might have caused this **change**, and in particular, the **scale** of this **change**. Estimate the proportion of this **environment** that has been affected. What proportion do you think is the result of human activity? Compare your estimate with the figure of 40 per cent given in section 2.4.3.

   Classifying, organising, constructing

3. Use the internet to investigate soils found in desert and rainforest **environments**. Construct a soil profile for each **place** and highlight the differences between them. Find out if the percentages shown in **FIGURE 4** are different in each **place**, and add this information to your soil profiles. **Classifying, organising, constructing**

4. Use the **Soil formation** weblink in the Resources tab. Describe the main steps in the formation of soil.

   Examining, analysing, interpreting

5. Think about what you have learned about soil formation.
   (a) Dig a hole outside where the soil has not been disturbed too much. Dig until you find small pieces of weathering rock. Measure the depth of your hole. How does this compare with the depth of soils in Australia and overseas?
   (b) Find two pieces of rock that show signs of weathering. Check the hardness of these rocks; the harder the rocks, the more difficult it will be to obtain a sample. Rub them together over a piece of paper. Were you able to collect a spoonful of grains in a reasonable amount of time? If so, how long did it take to rub off a spoonful of particles?
   (c) The rate of soil formation is estimated at less than 0.05 millimetres a year in eastern Australia. How long would it take to develop one centimetre of soil? How long would it take to form enough soil to replace what was in the hole you dug earlier?
   (d) Write a paragraph explaining what this exercise tells you about soil formation and the need to use soil in a **sustainable** manner.

   Examining, analysing, interpreting

2.4 EXERCISES

Geographical skills key: GS1 Remembering and understanding **GS2** Describing and explaining **GS3** Comparing and contrasting **GS4** Classifying, organising, constructing **GS5** Examining, analysing, interpreting **GS6** Evaluating, predicting, proposing

2.4 Exercise 1: Check your understanding

1. **GS1** What is soil?
2. **GS1** Why is soil important?
3. **GS2** In your own words, define the natural processes at work shaping the Earth.
4. **GS2** Explain the difference between weathering and erosion.
5. **GS1** Identify human factors that might contribute to erosion.

2.4 Exercise 2: Apply your understanding

1. **GS2** Explain how and why human activity might contribute to weathering and erosion.
2. **GS2** Using terms such as *uplift, erosion, deposition, weathering* and *transportation*, explain the **interconnection** between physical processes and the **environment**.
3. **GS2** In your own words, explain how soil is formed and why it is not uniform across the surface of the Earth.
4. **GS2** Using examples, describe two different ways that mountain ranges can be formed.
5. **GS5** Australia is an ancient landmass. Which processes described in this subtopic are currently shaping Australia's landforms? Justify your answer.

Try these questions in learnON for instant, corrective feedback. Go to www.jacplus.com.au.

2.5 Underground landscapes

2.5.1 What is karst?

Apart from rivers and streams that flow across the surface of the Earth, vast networks of rivers also exist under the ground. The result is a network of caves and channels that carve a very different landscape, known as karst.

Karst is a landscape formed by water dissolving bedrock (solid rock beneath soil) over hundreds of thousands of years (see **FIGURE 1**). On the surface of the Earth, sinkholes (holes in the Earth's surface), vertical shafts (tunnels), and fissures (cracks) will be evident. Rivers and streams may seem to simply disappear, but underground there are intricate drainage networks, complete with caves, rivers, **stalactites** and **stalagmites** (see **FIGURE 2**).

FIGURE 1 Formation of a karst landscape

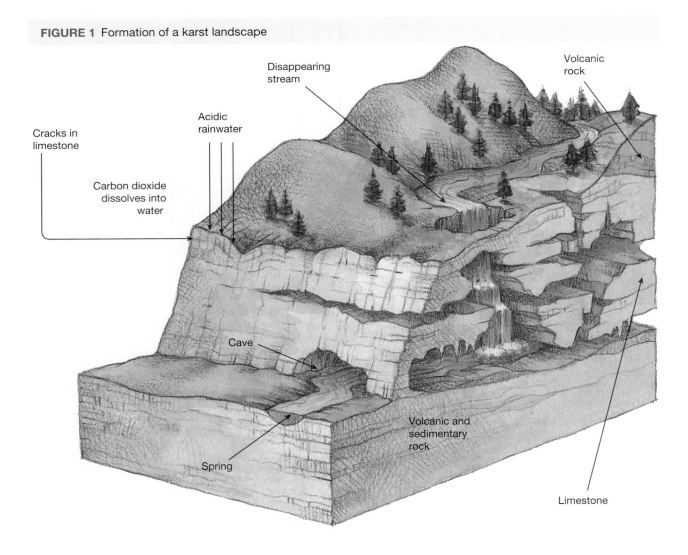

Karst topography makes up about 10 per cent of the Earth's surface; however, a quarter of the world's population depends on karst environments to meet its water needs.

2.5.2 How are karst landscapes formed?

Water becomes slightly acidic when it comes into contact with carbon dioxide in the atmosphere (as it does when raindrops form) or when it filters through organic matter in the soil and percolates into the ground. Acidic water is able to dissolve soluble bedrock, such as limestone and dolomite. This creates cracks or fissures, allowing more water to penetrate the rocks. When the water reaches a layer of non-dissolving rocks, it begins to erode sideways, forming an underground river or stream. As the process continues, the water creates hollows, eventually creating a cave. Some karst landscapes contain aquifers that are capable of providing large amounts of water.

FIGURE 2 Caves in Guilin, Guangxi Province, southern China

2.5.3 Where are karst landscapes found?

Karst landscapes are found all over the world, as shown in **FIGURE 3**, in locations where mildly acidic water is able to dissolve **soluble** bedrock such as limestone and dolomite.

In tropical regions, where rainfall is very high, karst mountains sometimes develop. This is because the high rainfall levels wear away the soluble rock much faster than rock is worn away in karst areas with lower rainfall. Examples of tropical karst mountains include the peaks of Ha Long Bay in Vietnam and the Guilin Mountains in China.

FIGURE 3 Karst regions of the world

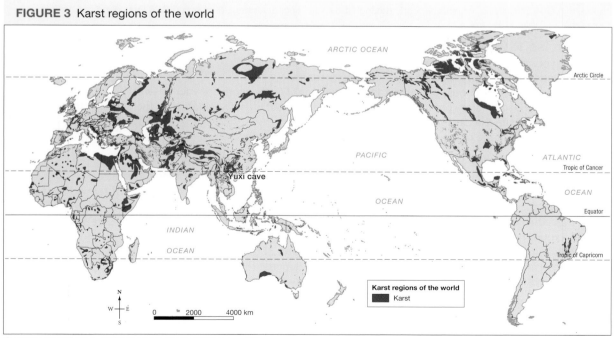

Source: World Map of Carbonate Rock Outcrops v3.0.

The Earth's largest arid limestone karst cave system is located on Australia's Nullarbor Plain, covering 270 000 square kilometres. It extends 2000 kilometres from the Eyre Peninsula in South Australia to Norseman in the Goldfields–Esperance region of Western Australia, and from the Bunda Cliffs on the Great Australian Bight in the south to the Victoria Desert in the north. The extensive cave system provides a unique habitat for a variety of native flora and fauna. Within the caves are fossils that can reveal much about our distant past, along with important Indigenous heritage sites.

 Resources

 Interactivity Underground wonders (int-3103)

2.5 INQUIRY ACTIVITIES

1. Examples of karst landscapes in Australia include the Buchan, Naracoorte, Jenolan, Labertouche, Princess Margaret Rose, Judbarra and Abercrombie caves. Working with a partner, investigate one of these *environments* and prepare an annotated visual display. Show its location on a map, and include the *scale*, features, land use and any concerns or threats to the *environment*. Include information on what is being done to ensure the *sustainable* management of the *place*. Share your findings with the rest of the class.

   Classifying, organising, constructing

2. The Nullarbor Plain cave *environment* is a popular destination for caving groups. Use the internet to investigate this *environment* and why people are attracted to it. Compare this *environment* to the one you studied in activity 1. Pay particular attention to the *scale* and *change* that has occurred in each *place*. Is one more fragile than the other? Explain. Suggest strategies for the *sustainable* management of karst in the Nullarbor.

   Examining, analysing, interpreting

2.5 EXERCISES

Geographical skills key: GS1 Remembering and understanding **GS2** Describing and explaining **GS3** Comparing and contrasting **GS4** Classifying, organising, constructing **GS5** Examining, analysing, interpreting **GS6** Evaluating, predicting, proposing

2.5 Exercise 1: Check your understanding

1. **GS2** In your own words, explain how a karst landscape is formed.
2. **GS2** Describe the global distribution of karst landscapes.
3. **GS2** Do you think we should preserve karst landscapes? Give reasons for your answer.
4. **GS1** Karst landscapes are predominantly found underground. Identify evidence on the surface of the Earth that might indicate the existence of a karst landscape.
5. **GS1** What percentage of the Earth's topography could be described as karst?

2.5 Exercise 2: Apply your understanding

1. **GS6** The world's largest arid limestone karst system is found on the Nullarbor Plain, Australia.
   (a) The Nullarbor Plain is an example of a desert landscape; suggest how an *environment* formed by water can occur in this location.
   (b) Describe how you think this landscape would be different if it were located in Australia's tropical north.
2. **GS5** Explain how the karst landscape can provide us with a link to our distant past.
3. **GS2** Explain how the karst landscape can provide a quarter of the world's population with water.
4. **GS6** Karst is often described as 'a hidden landscape'. Suggest reasons for this description.
5. **GS6** Suggest a reason for the absence of karst landscapes in Antarctica.

Try these questions in learnON for instant, corrective feedback. Go to www.jacplus.com.au.

2.6 Australian landforms

2.6.1 What processes have shaped Australasia?

The tectonic forces of folding, faulting and volcanic activity have created many of Australia's major landforms. Other forces that work on the surface of Australia, and give our landforms their present appearance, are weathering, mass movement, erosion and deposition.

Australia is an ancient landmass. The Earth is about 4600 million years old, and parts of the Australian continent are about 4300 million years old.

Over millions of years, Australia has undergone many changes. Mountain ranges and seas have come and gone. As mountain ranges eroded, sediments many kilometres thick were laid down over vast areas. These sedimentary rocks were then subjected to folding, faulting and uplifting. This means that the rocks that make up the Earth's crust have buckled and folded along areas of weakness, known as faults. Sometimes, fractures or breaks occur, and forces deep within the Earth cause sections to be raised, or uplifted. Over time the forces of weathering and erosion have worn these down again. Erosion acts more quickly on softer rocks, forming valleys and bays. Harder rocks remain as mountains, hills and coastal headlands.

Because it is located in the centre of a **tectonic plate**, rather than at the edge of one, Australia currently has no active volcanoes on its mainland, and has very little tectonic lift from below.

FIGURE 1 Many of Queensland's mountain peaks were formed by volcanic activity around 20 million years ago. The Glasshouse Mountains, north of Brisbane, are volcanic plugs. They are composed of volcanic rock that hardened in the vent of a volcano. Over millions of years, weathering and erosion have worn away the softer rock that surrounded the vent, leaving only the plugs.

This means its raised landforms such as mountains have been exposed to weathering forces for longer than mountains on other continents and are therefore more worn down.

About 33 million years ago, when Australia was drifting northwards after splitting from Antarctica, the continent passed over a large **hotspot**. Over the next 27 million years, about 30 volcanoes erupted while they were over the hotspot. The oldest eruption was 35 million years ago at Cape Hillsborough, in Queensland, and the most recent was at Macedon in Victoria around six million years ago. Over millions of years, these eruptions formed a chain of volcanoes in eastern and south-eastern Australia, that are known today as the Great Dividing Range (see **FIGURE 2**). At present, the hotspot that caused the earlier eruptions is probably beneath Bass Strait.

The present topography of much of Australia results from erosion caused by ice. For example, about 290 million years ago a huge icecap covered parts of Australia. After the ice melted, parts of the continent subsided and were covered by **sediment**, forming sedimentary basins (a low area where sediments accumulate) such as the Great Artesian Basin. On a smaller scale, parts of the Australian Alps and Tasmania have also been eroded by glaciers during the last ice age.

Rivers and streams are another cause of erosion, having carved many of the valleys in Australia's higher regions.

When streams, glaciers and winds slow down, they deposit or drop the material they have been carrying. This is called deposition. Many broad coastal and low-lying inland valleys have been created by stream deposition. These areas are called floodplains.

2.6.2 Australia's landform regions

The topography of Australia can be divided into four major regions (see **FIGURE 3**).

- The coastal lowlands around Australia's edge are narrow and fragmented. The plains often take the form of river valleys, such as the Hunter River Valley near Newcastle.
- The eastern highlands region (which includes the Great Dividing Range) is mainly a series of tablelands and plateaus. Most of the area is very rugged, because rivers have cut deep valleys. It is the source of most of Australia's largest rivers, including the Fitzroy, Darling and Murray. The highest part is in the south-east, where a small alpine area is snow-covered for more than half the year.
- The central lowlands are a vast area of very flat, low-lying land that contains three large **drainage basins**: the Carpentaria Lowlands in the north, the Lake Eyre Basin in the centre (see **FIGURE 4**) and the Murray–Darling Basin in the south.
- The Great Western Plateau is a huge area of tablelands, most of which are about 500 metres above sea level. It includes areas of gibber (or stony) desert and sandy desert. There are several rugged upland areas, including the Kimberley and the MacDonnell Ranges.

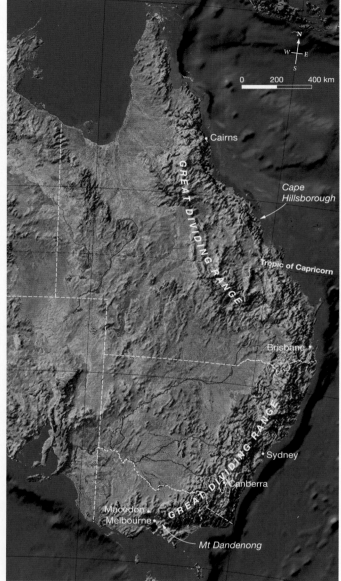

FIGURE 2 Relief map of Australia's east coast. The Great Dividing Range stretches from north of Cairns in Queensland to Mount Dandenong near Melbourne in the south.

Source: Spatial Vision

FIGURE 3 Australia's four major landform regions

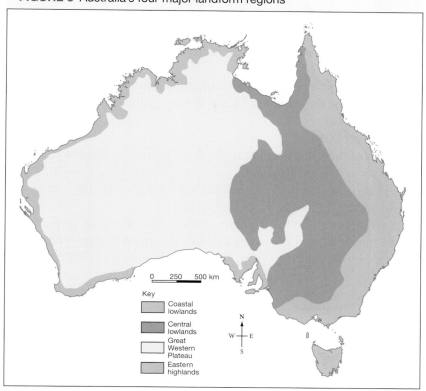

Key

Coastal lowlands

Central lowlands

Great Western Plateau

Eastern highlands

Source: MAPgraphics Pty Ltd Brisbane

FIGURE 4 Kati Thanda–Lake Eyre, the lowest point on the Australian mainland, is part of the Great Artesian Basin. It is 15 metres below sea level. Once a freshwater lake, the region is now the world's largest salt pan. The evaporated salt crust shows white in the satellite image (a) below. The lake fills with water only three or four times each century, transforming it into a haven for wildlife. Deep water is shown as black in image (b) below.

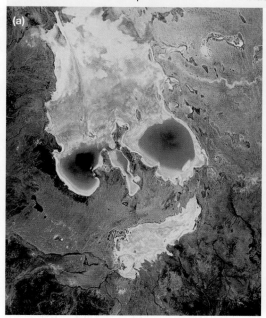

2.6.3 CASE STUDY: Water issues in the Murray–Darling Basin region

The Murray–Darling Basin covers about one million square kilometres, and more than 20 major rivers flow into it. It has a wide variety of landscapes, ranging from alpine areas in the south-east to plains in the west. The basin produces 43 per cent of Australia's food and over 40 per cent of Australia's total agricultural income.

The Murray–Darling Basin is the largest and most important drainage basin in Australia, covering one-seventh of the continent. However, the amount of water flowing through it in one year is about the same as the *daily* flow of the Amazon River.

The basin is facing severe problems.

- Only about 20 per cent of the water flowing through the basin ever reaches the sea. The rest is diverted for agriculture, industry and domestic use.
- The Murray supplies about 40 per cent of Adelaide's drinking water. The quality of the water continues to decline, mainly because of salinity levels.
- Approximately 50 to 80 per cent of the wetlands in the basin have been severely damaged or destroyed, and more than a third of the native fish species are threatened with extinction.
- River system inflows vary from year to year. The long-term average is 9030 GL. In 2018, inflows were around 2740 GL, among the lowest on record.
- An estimate of weather trends shows that the flow to the Murray River mouth may be reduced by a further 25 per cent by 2030. However, with the added problem of climate change, it is predicted that precipitation in the Murray–Darling catchment will decrease, so that the reduction in flow to the mouth could be as high as 70 per cent.

FIGURE 5 Aerial view of the Murray River, where it enters the Coorong and Lake Alexandrina in South Australia

Explore more with my**World**Atlas

Deepen your understanding of this topic with related case studies and questions.
- Investigate additional topics > Managing water resources > **Murray–Darling Basin**

2.6.4 How does water flow across the land?

Permanent rivers and streams flow in only a small proportion of the Australian continent. Australia is in fact the driest of all the world's inhabited continents. It has:
- the least amount of run-off
- the lowest percentage of rainfall as run-off
- the least amount of water in rivers
- the smallest area of permanent wetlands
- the most variable rainfall and stream flow.

FIGURE 6 Australia's drainage basins

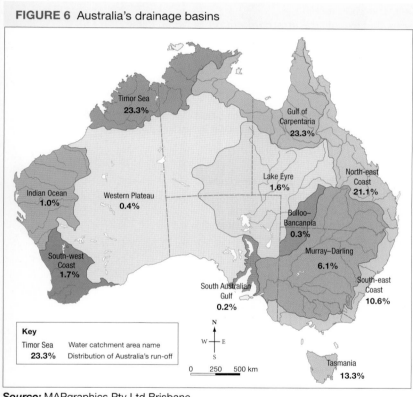

Source: MAPgraphics Pty Ltd Brisbane

FIGURE 7 Kati Thanda–Lake Eyre and surrounding drainage systems

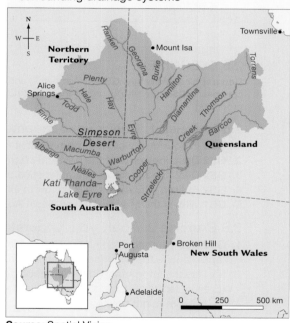

Source: Spatial Vision

Australia has many lakes, but they hold little water compared with those found on other continents. The largest lakes are Kati Thanda–Lake Eyre (see **FIGURES 6** and **7**) and Lake Torrens in South Australia. During the dry seasons, these become beds of salt and mud. Yet an inland sea did once exist in this area. It covered about 100 000 square kilometres around present-day Kati Thanda–Lake Eyre and Lake Frome. South Australia is Australia's driest state, and has very few permanent rivers and streams.

Explore more with my**World**Atlas

Deepen your understanding of this topic with related case studies and questions.
- Investigating Australian Curriculum topics > Year 8: Landforms and landscapes > **Uluru**

2.6 INQUIRY ACTIVITIES

1. Use your atlas to list the highest mountains in each Australian state and territory. Describe the location of each. **Describing and explaining**
2. Use Google Earth to view any part of the Murray–Darling Basin. Describe the landscape that you see. **Describing and explaining**
3. Divide your class into four groups. Assign each group one of Australia's landform regions to investigate. Collectively compile a list of landforms that are found in each region. Then have each member of the group investigate a different landform and prepare a series of PowerPoint slides that show the following:
   (a) the landform
   (b) where it is located
   (c) how it was formed
   (d) whether people might want to visit this landform, including the reasons why it is or is not a popular landform.
   Put the individual presentations together for viewing by the rest of the class. **Describing and explaining**
4. Australia is an ancient landmass and has undergone many *changes* over millions of years. In groups, brainstorm and compile lists under the following headings.
   • Physical *changes* that have taken place on the Australian landmass
   • Tectonic processes that have contributed to these *changes*
   • *Changes* caused by processes such as weathering and erosion
   Within your group, write a series of paragraphs that explain the *interconnection* between these factors. **Examining, analysing, interpreting**

2.6 EXERCISES

Geographical skills key: GS1 Remembering and understanding **GS2** Describing and explaining **GS3** Comparing and contrasting **GS4** Classifying, organising, constructing **GS5** Examining, analysing, interpreting **GS6** Evaluating, predicting, proposing

2.6 Exercise 1: Check your understanding

1. **GS1** In your own words, explain what is meant by the terms *folding, faulting* and *uplift*.
2. **GS2** Describe some of the physical *changes* Australia's landmass has undergone.
3. **GS2** Describe the major characteristics of Australia's four main landform regions.
4. **GS2** Explain why Australia is so low in altitude and flat compared with other continents.
5. **GS1** Why is the Murray–Darling Basin Australia's most important drainage basin?

2.6 Exercise 2: Apply your understanding

1. **GS6** Use your atlas to find the Cape Hillsborough and Macedon volcanoes, or refer to **FIGURE 2**.
   (a) Calculate the distance between them.
   (b) Use the information in this subtopic to work out the rate at which the Australian landmass is moving.
   (c) How far has Australia moved over the Bass Strait hotspot? Now calculate where under Bass Strait this hotspot might now lie.
   (d) Use the information in this subtopic to explain why this hotspot has *changed* its location over time.
2. **GS3** It is said that the amount of water that flows down the Amazon River in a day is more than flows down the Murray in a year.
   (a) What does that tell you about how dry Australia's climate is?
   (b) How might this affect the *environment* around the Murray River?
3. **GS2** Describe the role of the Bass Strait hotspot in creating the landforms on Australia's east coast.
4. **GS6** Describe how Kati Thanda–Lake Eyre has *changed* over time. Suggest a reason for these *changes*.
5. **GS6** Approximately 80 percent of the water flowing through the Murray–Darling Basin is diverted.
   (a) What is this water used for?
   (b) What impact might this have on people and the *environment*?

Try these questions in learnON for instant, corrective feedback. Go to www.jacplus.com.au.

2.7 Landforms of the Pacific

2.7.1 What is the Pacific landscape like?

The Pacific Ocean is the world's largest ocean, and occupies almost a third of the Earth's surface, making it larger than all the Earth's land areas combined. It stretches from the Arctic in the north to Antarctica in the south and is bordered by Australia and Asia in the west and the Americas in the east. The 25 000 Pacific Islands of Polynesia, Micronesia and Melanesia (including Papua New Guinea) are home to around 10 million people.

The Pacific islands are broken up into three main island groups (see **FIGURE 1**).

FIGURE 1 The Pacific Ocean

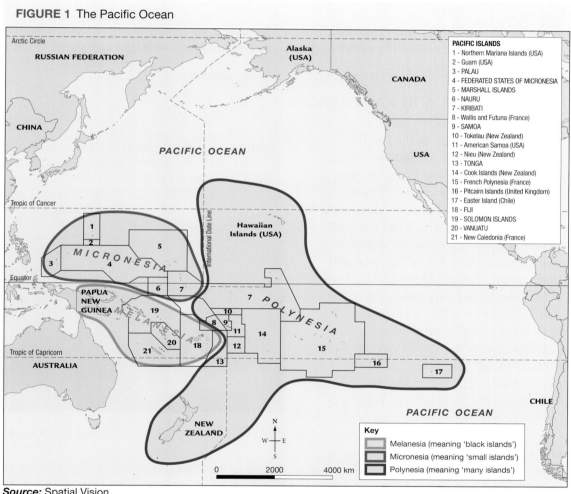

Source: Spatial Vision

Melanesia extends north and north-east of Australia, from the west Pacific Ocean to the Arafura Sea. It includes the islands of New Guinea (the nation of Papua New Guinea and the Indonesian province of Papua), New Caledonia, Vanuatu, Fiji and the Solomon Islands.

Micronesia has hundreds of small islands and is located north-east of Papua New Guinea. It is bounded by the Philippines in the west, Indonesia in the south-west and Melanesia to the south. Micronesia includes the Northern Marianas, Guam, Palau, the Marshall Islands, Kiribati, Nauru and the Federated States of Micronesia.

Polynesia forms a triangle, with its three corners at Hawaii, New Zealand and Easter Island. There are around 1000 islands in this part of the central southern Pacific Ocean. The other main islands are Samoa, Tonga, French Polynesia, Tuvalu and the Cook Islands.

Amazing Pacific facts

Apart from being the world's largest ocean, the Pacific holds a number of other records.

- At 10 203 metres, Mauna Kea in Hawaii is the highest mountain from base to summit.
- Mauna Loa is the world's largest active volcano. It is 120 kilometres long and 50 kilometres wide. It has been active for over 700 000 years and will most likely continue to erupt for another 500 000 years.
- The Mariana Trench in the western Pacific is the deepest point on Earth — 11 032 metres.
- Australia's Great Barrier Reef is the world's largest coral reef, stretching some 2027 kilometres.
- Kwajalein in the Marshall Islands, with a length of 125 kilometres, is the world's largest **coral atoll**. It actually comprises 97 islands and **islets** and surrounds one of the world's largest **lagoons**, covering 2173 square kilometres.
- Grand Lagoon Sud in New Caledonia is the world's largest lagoon, covering an area of 3145 square kilometres.
- The Pacific Ocean is encircled by the Pacific Ring of Fire, the world's most active tectonic region. Approximately 75 per cent of the world's active volcanoes are here and 90 per cent of the world's earthquakes occur in this region.

2.7.2 High islands and low islands

What are high islands?

A number of the high islands in the Pacific were once part of either the Australian or Asian continents. These include New Zealand, New Guinea and most of the islands in Melanesia.

Other high islands are volcanic and are really the tops of undersea mountains. They are made up of magma that was forced up through fissures (cracks) in the ocean floor before being cooled by sea water and hardening. Many of the islands found in Micronesia and Polynesia were formed in this way. Sometimes volcanic islands are formed in a chain called an archipelago.

New Zealand has more than 200 islands and 220 mountains higher than 2300 metres, the highest being Mount Cook at 3754 metres.

New Guinea is also mountainous. It has a central spine formed by high mountain ranges, the

FIGURE 2 An underwater volcanic eruption in 2009 created a new island off the coast of Tonga, an island group in the South Pacific.

highest of which is Puncak Jaya at 5030 metres. Tahiti, Fiji, Vanuatu, the Caroline Islands and Raratonga in the Cook Islands are also volcanic islands.

Many large rivers flow from these high mountains, including the Fly River in Papua New Guinea, the Waikato in New Zealand, and the Rewa and Sigatoka Rivers on Viti Levu in Fiji. Many high islands have fertile volcanic soils that support a variety of vegetation types, including rainforest, mangrove forests and palms.

What are low islands?

Some of the low islands in the Pacific are the remains of volcanoes that have eroded over time and are now only just above sea level. Examples include some of the smaller islands in Hawaii and Bora Bora in French Polynesia.

Other low islands are reefs, or atolls built on coral reefs, and are usually quite small, some barely reaching above sea level. Low islands are often a series of very small islands and islets with a lagoon at their centre, known as atolls. Some coral atolls were built by volcanic activity millions of years ago. Eventually, when the volcanic island erodes, it leaves a lagoon in its place.

Micronesia and Polynesia are dominated by low islands. Mount Orohena and Mount Aorai on the island of Tahiti and Mount Tohiea on the island of Moorea (both in French Polynesia) are examples of old volcanoes. They have fringing reefs surrounding a shallow lagoon formed when the island was eroded. These reefs have become barrier reefs that protect the island and the lagoon from the force of the ocean waves.

Coral atolls have no rivers, and the soil is generally thin and not very fertile. Some low islands receive high rainfall, and have more fertile soils that can sometimes support forests.

FIGURE 3 The volcanic island of Moorea, French Polynesia, surrounded by a fringing reef

FIGURE 4 Development of lagoons and fringing reefs

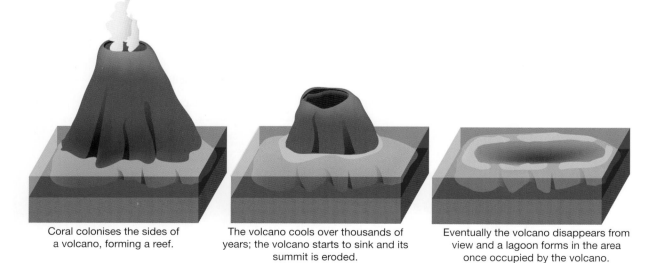

Coral colonises the sides of a volcano, forming a reef.

The volcano cools over thousands of years; the volcano starts to sink and its summit is eroded.

Eventually the volcano disappears from view and a lagoon forms in the area once occupied by the volcano.

Aboriginal peoples and Torres Strait Islander peoples are recognised as the first Australians. Evidence of their presence in Australia is found across the continent in their rock art (as shown in **FIGURE 1**), in **archaeological** records, and through their cultural heritage passed down through generations. As **hunter–gatherers** they relied on the plants, animals and the environment for their survival, and so have an understanding of the complex nature of Australia's varying landscape.

Europeans, on the other hand, arrived in 1788 and occupied areas of Australia. They had a very different view of the landscape, based on ideas they brought with them from Britain. They sought to change the landscape and adapt it to meet their needs. They established permanent settlements and depended on agriculture to provide for their needs.

The perspective of Indigenous Australian peoples is one of being part of the landscape, while the European perspective is based on the idea of land ownership.

2.9.2 Kakadu — Australia's first World Heritage Area

Kakadu National Park, as seen in **FIGURE 2**, covers an area of approximately 20 000 square kilometres of the Northern Territory — an area roughly a third the size of Tasmania. It stretches 200 kilometres from north to south, and spans 100 kilometres from east to west. Within the boundaries of the park are vast uranium deposits. Kakadu is unique in that it is recognised for both its natural beauty and its cultural value.

FIGURE 2 Map of Kakadu National Park

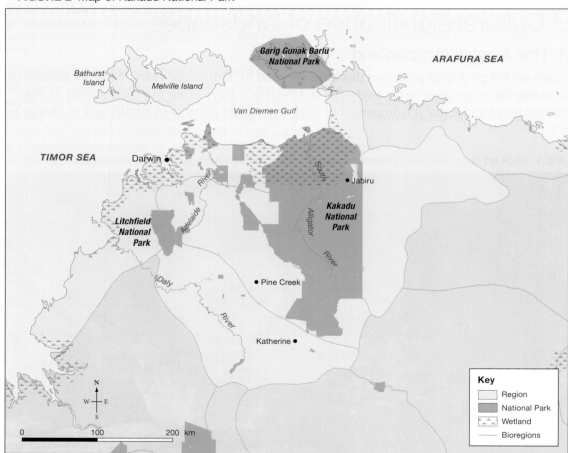

Source: Spatial Vision

FIGURE 3 Why Kakadu is valued

2.9.3 Kakadu and its resources

Kakadu is rich with the historical records and ancestry of the first Australians. In addition, it supports a treasure trove of native plant and animal species and provides a temporary home to a large number of migratory birds. More than 200 000 tourists visit Kakadu annually, attracted by its vast wetlands and scenery, including steep gorges, Aboriginal rock art, lookouts, and waterfalls such as Jim Jim Falls (see **FIGURE 4**).

Kakadu also has vast deposits of uranium ore, which is a potentially valuable export for Australia. Opponents of uranium mining are concerned about the possibility that Australia's uranium could be processed and used to make nuclear weapons. Others fear the effects of mining on the environment and the potential for a devastating pollution event.

The Ranger uranium mine has been operating since 1980 and lies within the boundaries of Kakadu National Park. Three kilometres downstream from the mine, the Mirrar people (a local Aboriginal community) swim and fish. Since the mine opened, there have been more than 200 leaks and spills, and the mine has generated some 30 million tonnes of liquid radioactive waste (see **FIGURE 5**). The mine is scheduled for closure in 2021. Some parts of the mine area are undergoing rehabilitation, with an extensive $800 million rehabilitation program scheduled for when the mine finally closes.

FIGURE 4 Jim Jim Falls at Kakadu is a popular tourist destination.

FIGURE 5 Timeline of major breaches at the Ranger uranium mine since 2002

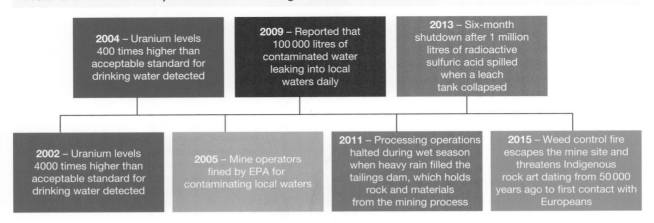

2004 – Uranium levels 400 times higher than acceptable standard for drinking water detected

2009 – Reported that 100 000 litres of contaminated water leaking into local waters daily

2013 – Six-month shutdown after 1 million litres of radioactive sulfuric acid spilled when a leach tank collapsed

2002 – Uranium levels 4000 times higher than acceptable standard for drinking water detected

2005 – Mine operators fined by EPA for contaminating local waters

2011 – Processing operations halted during wet season when heavy rain filled the tailings dam, which holds rock and materials from the mining process

2015 – Weed control fire escapes the mine site and threatens Indigenous rock art dating from 50 000 years ago to first contact with Europeans

DISCUSS
Why do you think the Australian government allows uranium mining in such an important region of Australia?
[Ethical Capability]

2.9 INQUIRY ACTIVITY

Write a letter to the editor of a newspaper outlining your views on uranium mining in *environmentally* sensitive areas. Explain whether you consider this type of activity a *sustainable* use of the landscape.

Describing and explaining

2.9 EXERCISES

Geographical skills key: GS1 Remembering and understanding **GS2** Describing and explaining **GS3** Comparing and contrasting **GS4** Classifying, organising, constructing **GS5** Examining, analysing, interpreting **GS6** Evaluating, predicting, proposing

2.9 Exercise 1: Check your understanding

1. **GS1** Where is Kakadu National Park and why is it important?
2. **GS3** Copy the table below into your workbook and use it to compile a list of differences in the way the Australian landscape was viewed by Aboriginal and Torres Strait Islander peoples and non-Aboriginal and Torres Strait Islander Australians. The first one has been done for you.

Aboriginal and Torres Strait Islander peoples views	Non-Aboriginal and Torres Strait Islander views
The land is communally owned.	Individuals own the land.

3. **GS2** Consider the Indigenous Australian population.
   (a) Where are the more densely populated regions of Australia? *Hint:* Find a map in your atlas that shows population distribution.
   (b) Why would it be more difficult for Aboriginal and Torres Strait Islander communities in these areas to maintain their traditional lifestyle and culture?
4. **GS2** Describe the *interconnection* that Aboriginal and Torres Strait Islander peoples have with the landscape. What evidence of this *interconnection* is found in this subtopic?
5. **GS1** Consider the resources in the Kakadu region.
   (a) What is uranium used for and why is it considered a valuable resource?
   (b) What risks does uranium mining pose in the Kakadu region?

2.9 Exercise 2: Apply your understanding

1. **GS5** Think about your personal values and beliefs and analyse how they might be similar or different to those reflected in **FIGURE 3**.
2. **GS6** Think back to the section on mining in the Kakadu region.
   (a) Suggest three possible impacts on the landscape if a new uranium mine was opened in the Kakadu region.
   (b) Do you think *changes* would have a large-*scale* or a small-*scale* impact? Explain.
3. **GS6** Predict what pressures decision makers in Australia might face in future when balancing the needs of the different groups who have an interest in Kakadu's resources.
4. **GS2** Australia's first people did not have a written language. Explain how we have such an extensive knowledge of their culture, history and beliefs.
5. **GS6** Present one argument for and one argument against granting leases to mine resources such as uranium in the Kakadu region.

Try these questions in learnON for instant, corrective feedback. Go to www.jacplus.com.au.

2.10 Preserving and managing landscapes

2.10.1 The World Heritage Convention

Worldwide, people recognise the value of landscapes and the need to protect their natural beauty and cultural heritage, and to manage their resources sustainably. Landscapes are easily damaged or destroyed but are difficult to recreate and repair. The key is to ensure that they are carefully managed so that the landscapes we value today are still present in the future.

From the middle of the twentieth century, there was growing concern about the need to protect areas of both cultural and natural significance (see **FIGURE 1**).

FIGURE 1 The World Heritage list includes 1092 sites of significance.

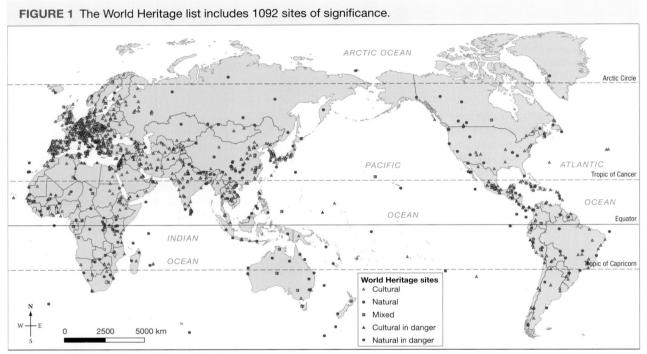

DISCUSS

Some natural landscapes can be loved to death when they are visited and used by large numbers of people, sometimes having a negative impact on the landscape. As a class, discuss how several Victorian landscapes could be managed and whether people should be allowed to use them.

[Critical and Creative Thinking Capability]

2.10.2 The Artesian Range

The Artesian Range is a unique part of the Australian landscape. It has been described as a lost world, a modern-day Noah's Ark, our last opportunity to protect and preserve a part of the Australian mainland that has had little contact with modern civilisation. Within its hidden valleys and canyons lies a diverse range of flora and fauna. The rich tropical rainforests and woodlands provide vital habitats for some of Australia's most endangered wildlife.

The Artesian Range covers 1800 square kilometres (see **FIGURE 2**). It is largely inaccessible; the only way in is by helicopter or boat. It is a maze of hidden valleys and canyons, rocky ranges and plateaus, towering **escarpments**, wide valleys and deep gorges (see **FIGURE 3**). Its sandstone ranges were formed as a result of tectonic plate activity. These rock formations date back some 1.8 million years.

Although it is difficult for humans to reach the area, exotic species such as donkeys, horses, pigs and cats have gradually invaded the Kimberley. And while fire is a natural part of the landscape, changing fire patterns and the increasing number of late-season wildfires are also a threat to the Artesian Range. Australian Wildlife Conservancy (AWC), an independent non-profit organisation funded by donations has now secured the land and manages it for conservation. AWC undertakes fire management, feral animal control, and biological surveys and monitoring, protecting the full length of the Artesian Range.

FIGURE 2 The Artesian Range covers 1800 square kilometres of the Kimberley region.

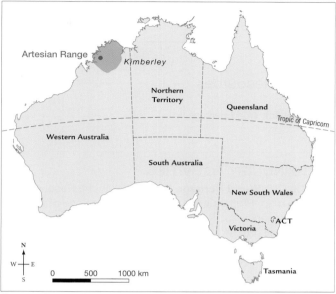

Source: Spatial Vision

FIGURE 3 The Artesian Range is a rugged and largely inaccessible landscape, renowned for its natural beauty and unique wildlife.

Source: AWC/Wayne Lawler

on Resources

🔗 **Weblink** World Heritage list

2.10 INQUIRY ACTIVITIES

1. Use the **World Heritage list** weblink in the Resources tab and select a site in one of the countries listed on the map. Prepare a visual presentation of one of the sites listed, outlining its importance and how it is protected. **Classifying, organising, constructing**

2. In small groups, investigate an invasive species and describe the ways in which it has **changed** the **environment**. Is this **change** occurring on a small or large **scale**? Explain. Suggest a strategy that the Australian Wildlife Conservancy could employ to eradicate invasive species from this **environment**. **Evaluating, predicting, proposing**

3. (a) Explain what you understand by the terms *cultural significance* and *natural significance*.
   (b) Is it possible for **places** to have both cultural and natural significance? Draw up a table like the one below. With the aid of a partner, add as many **places** as you can to the list. Try to have a balance of Australian and international examples. Compare your list with that of another pair of students.

Cultural significance	Natural significance	Cultural and natural significance

   (c) Which column has the most entries? Suggest a reason for the pattern you observe.
   (d) Select one **place** from column 3. Find a picture of this **place** and copy and paste it into a Word document. Add annotations to explain the major features of your chosen **place** and why it is of cultural and natural significance. **Examining, analysing, interpreting**

2.10 EXERCISES

Geographical skills key: GS1 Remembering and understanding **GS2** Describing and explaining **GS3** Comparing and contrasting **GS4** Classifying, organising, constructing **GS5** Examining, analysing, interpreting **GS6** Evaluating, predicting, proposing

2.10 Exercise 1: Check your understanding

1. **GS2** Why is it important to protect sites that have cultural or natural significance?
2. **GS2** Describe the location of the Artesian Range and why it is unique.
3. **GS2** Suggest why the Artesian Range has been largely inaccessible to people.
4. **GS1** Identify two management strategies used by the Australian Wildlife Conservancy (AWC) to manage and conserve the Artesian Range.
5. **GS6** Do you think invasive species or wildfires pose the greatest risk to the Artesian Range? Give a reason for your answer.

2.10 Exercise 2: Apply your understanding

1. **GS2** Explain how exotic species such as cats, foxes and camels have been able to become established in the Artesian Range when it is difficult for people to enter the region.
2. **GS6** Evaluate the ways in which the community demonstrates the value it places on cultural diversity and why this is important to the community.
3. **GS5** The Artesian Range has been described as a 'modern-day Noah's Ark'. Explain what you understand by this description.
4. **GS2** Describe the processes that have led to the formation of the Artesian Range and its different landscape features.
5. **GS6** Uluru is considered to have both cultural and natural significance. Suggest a reason for this classification.

Try these questions in learnON for instant, corrective feedback. Go to www.jacplus.com.au.

2.11 Thinking Big research project: Karst landscape virtual tour

SCENARIO

To acknowledge the importance of the karst environment and its connection to Indigenous Australian Dreaming stories, you are to create a karst landscape virtual tour for the National Museum.

Select your learnON format to access:

- the full project scenario
- details of the project task
- resources to guide your project work
- an assessment rubric.

 **Resources**

projectsPLUS Thinking Big research project: Karst landscape virtual tour (pro-0168)

2.12 Review

2.12.1 Key knowledge summary

Use this dot point summary to review the content covered in this topic.

2.12.2 Reflection

Reflect on your learning using the activities and resources provided.

 **Resources**

eWorkbook   Reflection (doc-31344)
            Crossword (doc-31345)

Interactivity   Introducing landforms and landscapes crossword (int-7595)

KEY TERMS

aquifer a body of permeable rock below the Earth's surface that contains water, known as groundwater

archaeological concerning the study of past civilisations and cultures by examining the evidence left behind, such as graves, tools, weapons, buildings and pottery

coral atoll a coral reef that partially or completely encircles a lagoon

deposition the laying down of material carried by rivers, wind, ice and ocean currents or waves

drainage basin an area of land that feeds a river with water; or the whole area of land drained by a river and its tributaries

erosion the wearing away and removal of soil and rock by natural elements, such as wind and water, and by human activity

escarpment a steep slope or long cliff formed by erosion or vertical movement of the Earth's crust along a fault line

glacier a large body of ice, formed by an accumulation of snow, which flows downhill under the pressure of its own weight

hotspot an area on the Earth's surface where the crust is quite thin, and volcanic activity can sometimes occur, even though it is not at a plate margin

hunter–gatherers people who collect wild plants and hunt wild animals rather than obtaining their food by growing crops or keeping domestic livestock

islet a very small island

lagoon a shallow body of water separated by islands or reefs from a larger body of water, such as a sea

mantle the layer of the Earth between the crust and the core

permafrost a layer beneath the surface of the soil where the ground is permanently frozen

plateau an extensive area of flat land that is higher than the land around it. Plateaus are sometimes referred to as tablelands.

sediment material carried by water

soluble able to be dissolved in water

stalactite a feature made of minerals, which forms from the ceiling of limestone caves, like an icicle. They are formed when water containing dissolved limestone drips from the roof of a cave, leaving a small amount of calcium carbonate behind.

stalagmite a feature made of minerals found on the floor of limestone caves. They are formed when water containing dissolved limestone deposits on the cave floor and builds up.

tectonic plate one of the slow-moving plates that make up the Earth's crust. Volcanoes and earthquakes often occur at the edges of plates.

transportation the movement of eroded materials to a new location by elements such as wind and water

weathering the breaking down of bare rock (mainly by water freezing and cooling as a result of temperature change) and the effects of climate

3 Landscapes formed by water

3.1 Overview

From gentle rain to rushing rivers, how does simple water form and transform landscapes?

3.1.1 Introduction

Water is one of the most powerful agents in creating landscapes. If you have ever been caught outside in a heavy downpour, walked through a fast-flowing creek, or been dumped in the surf, then you have felt and seen the energy of flowing water. It can knock you off your feet, move buildings and carve huge holes in the Earth's surface. Landscapes created by water are found everywhere.

 **Resources**

☑ **eWorkbook**   Customisable worksheets for this topic

▥ **Video eLesson** Landscapes sculpted by water (eles-1624)

To access a pre-test and starter questions and receive immediate, **corrective feedback** and **sample responses** to every question, select your learnON format at www.jacplus.com.au.

3.2 Landscapes formed by water

3.2.1 How does water change landscape features?

A torrent of gushing water can shift rocks, remove topsoil or shape river valleys. Gentle rain can change the chemical structure of any surface material, sculpting the imposing coastal landforms we see around the world. In cold climates, compressed snow in glaciers works like a slow-moving bulldozer to erode land and create unique landscape features. Once fresh water has made its way to the ocean, the power of waves creates coastal landscape features.

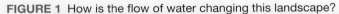

FIGURE 1 How is the flow of water changing this landscape?

As you learned in topic 2, landscapes are predominantly changed or created by two processes: **erosion** and **deposition**. Through erosion, water can carve through rock — reducing once-mighty cliffs to lowly sea-stacks. Through deposition, water creates beaches, spits and sand dunes as it carries sand across the oceans of the world. In **FIGURE 1** you can see the power of water as it rushes over a rockface and carves pools in its hard surface. You may have seen pools of a similar shape carved by waves in rocky coastal landforms.

As water makes contact with landscapes, it can change the shape and size of its features or landforms (**FIGURES 1** and **3**). The coastal landscape that you see today is not the same as it was hundreds or thousands of years ago. **FIGURE 2** is a photo of the Twelve Apostles, located on the south-western coast of Victoria. The name suggests that there may once have been twelve pillars of rock, or stacks, visible along this stretch of coastline. In the foreground you can see the remnants of two quite recently collapsed stacks. Even these stacks were once joined to the cliffs as part of the mainland. This highly erodible coastline has been constantly altered by many years of rainfall and wave action on the soft limestone cliffs.

FIGURE 2 The Twelve Apostles in Port Campbell National Park, Victoria. How might the potential for erosion change along this coast if the waves were larger and it was high tide?

Stacks show where the coastline used to be.

Current coastline

Collapsed stacks

Limestone cliffs

FIGURE 3 Water constantly moves over and through the Earth and through the air.

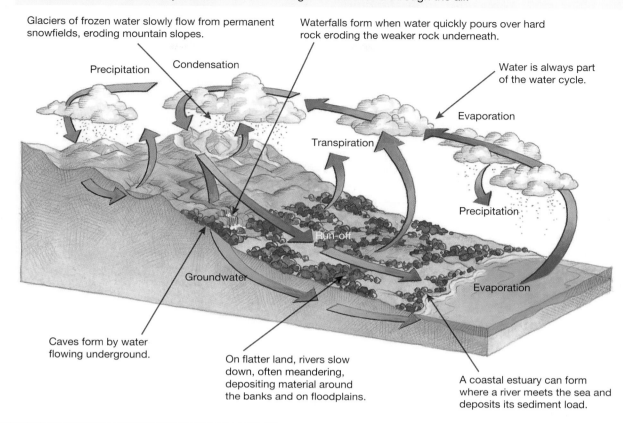

Glaciers of frozen water slowly flow from permanent snowfields, eroding mountain slopes.

Waterfalls form when water quickly pours over hard rock eroding the weaker rock underneath.

Water is always part of the water cycle.

Precipitation

Condensation

Evaporation

Transpiration

Precipitation

Run-off

Groundwater

Evaporation

Caves form by water flowing underground.

On flatter land, rivers slow down, often meandering, depositing material around the banks and on floodplains.

A coastal estuary can form where a river meets the sea and deposits its sediment load.

3.2 INQUIRY ACTIVITY

Use your research skills to create a list of world water facts on the following:
a. the biggest glacier
b. the longest river
c. the biggest wave
d. the highest waterfall
e. the widest river
f. the biggest ocean
g. a world water fact of your choice.
Show on a map where each is located.

Classifying, organising, constructing

3.2 EXERCISES

Geographical skills key: GS1 Remembering and understanding **GS2** Describing and explaining **GS3** Comparing and contrasting **GS4** Classifying, organising, constructing **GS5** Examining, analysing, interpreting **GS6** Evaluating, predicting, proposing

3.2 Exercise 1: Check your understanding

1. **GS1** Landscapes are in a state of continual **change**.
   (a) Which two natural processes powered by water are most responsible for continually **changing** landscapes?
   (b) How are these two processes linked?
2. **GS2** Where would **FIGURES 1** and **2** be **placed** on the landscape depicted in **FIGURE 3**? Explain.
3. **GS2** Explain how the water cycle and the formation of landscapes are **interconnected**.
4. **GS4** Draw your own copy of the diagram shown in **FIGURE 3**. Make sure that you included your own versions of the annotations as well.
5. **GS1** Of the two processes discussed in this subtopic, which is the most powerful — erosion or deposition?

3.2 Exercise 2: Apply your understanding

1. **GS2** Many landscapes *change* rapidly; for example, the Twelve Apostles.
   (a) Describe another example of a landscape that has been shaped by the power of water.
   (b) Do you think the *changes* to the landscape have been positive or negative?
   (c) To what extent should people try to stop the *changes* caused by water?
2. **GS2** Water can be considered one of the most important architects of desert landscape features. After looking at the images in this subtopic, try to explain how you think water can change the landscapes of arid or desert *environments*.
3. **GS6** Identify three possible ways that people can *change* the flow of water, either across the surface of the Earth or along the coast. Predict how you believe this may alter landscape features. Examples may include the use of river water for irrigation or the construction of a marina.
4. **GS2** Think back to the last time you visited a coastal *environment*. What features were prominent in the *environment* you visited? What processes were responsible for the creation of these features?
5. **GS6** Erosion and deposition are two processes that can transform coastal landscapes. Describe an additional way in which coastal landscapes can be *changed*.

Try these questions in learnON for instant, corrective feedback. Go to www.jacplus.com.au.

3.3 Coastal erosion

3.3.1 How do waves change an environment?

The coast is the zone or border between land and ocean. It is in this collision zone that the movement of sea water and the impact of the ocean on the land together create coastal landscapes. Coastal landscapes have landforms that are common to coastlines in different places around the world because they are built up or worn away in similar ways.

Before we investigate the different types of coastal landforms that exist, we need to first understand the processes which shape these landforms. Coastal erosion is mostly caused by the continued presence of waves in an environment. Waves are caused when the wind blows over the ocean. The size of a wave depends on the strength of the wind and the distance the wind has been blowing (referred to as the *fetch*). A strong wind and a long fetch will result in a powerful wave with a high degree of erosive potential. These waves are called **destructive waves** and they are involved in creating landforms by erosion. A gentle wind and a small fetch will create less powerful waves known as constructive waves. While these waves are not involved in erosion, they do create depositional landforms (see subtopic 3.4).

Next time you are walking along a beach, stop to check whether the waves in this environment are constructive or destructive. You can do this by analysing the strength of the **swash** and **backwash**. As a wave hits the shore it sends water (as well as sand, shells and other debris) onto the beach. This is called the swash. Water is then pulled back into the ocean by gravity in what is known as the backwash. If the swash is more powerful than the backwash, the waves are constructive and you should see depositional landforms. If the backwash is more powerful than the swash, the waves are destructive and you should see more landforms which have been caused by erosion. The structure of constructive and destructive waves can be seen in **FIGURE 1**.

Coastal landforms are not solely created by the power of waves. Rainfall and constant strong winds can also influence the appearance of coastal landforms. For example, after a puddle of rain water evaporates, it leaves behind salts and minerals which can interact with rocks. This can lead to scarring of the rock surface and, over time, deep crevasses can be formed. Other physical processes can also greatly affect the coastal landscape; for example, the tectonic force of earthquakes and volcanoes; changing sea levels; and human activities such as building roads, ports and houses, and damming rivers.

Explore more with myWorldAtlas

Deepen your understanding of this topic with related case studies and questions.
• Investigate additional topics > Oceans and coasts > **Coastal processes**

FIGURE 1 Comparing constructive and destructive waves

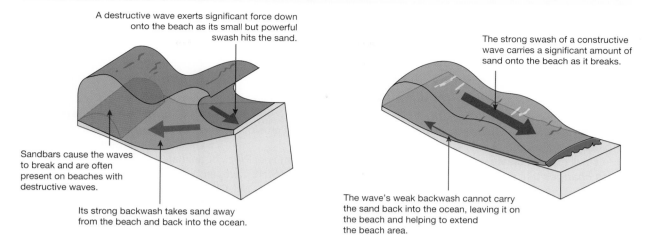

A destructive wave exerts significant force down onto the beach as its small but powerful swash hits the sand.

The strong swash of a constructive wave carries a significant amount of sand onto the beach as it breaks.

Sandbars cause the waves to break and are often present on beaches with destructive waves.

Its strong backwash takes sand away from the beach and back into the ocean.

The wave's weak backwash cannot carry the sand back into the ocean, leaving it on the beach and helping to extend the beach area.

Which coastal landscape features are created by erosion?

Features such as cliffs, headlands, bays, arches, caves, blow-holes and stacks are all landforms found along an eroding coastline (**FIGURE 2**). These features are formed by wave action and rainfall, which attack the cliffs and find points of weakness that are then eroded. Water running off a cliff face can carry eroded material into the sea below. When waves hit the cliff face, they undercut the base of the cliff to form a notch. As the notch increases in size it forms a cave and eventually the cliff gets undercut, becomes unstable and falls into the sea.

Destructive waves can also alter a sandy coastline. They can remove sand from a beach, destroy the vegetation on dunes, and remove management features designed to protect landscape features.

FIGURE 2 Coastal landforms created by erosion

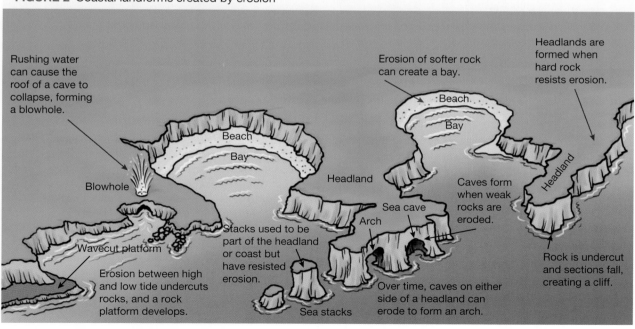

Rushing water can cause the roof of a cave to collapse, forming a blowhole.

Erosion of softer rock can create a bay.

Headlands are formed when hard rock resists erosion.

Beach

Bay

Beach

Bay

Blowhole

Headland

Caves form when weak rocks are eroded.

Headland

Wavecut platform

Sea cave

Arch

Erosion between high and low tide undercuts rocks, and a rock platform develops.

Stacks used to be part of the headland or coast but have resisted erosion.

Over time, caves on either side of a headland can erode to form an arch.

Rock is undercut and sections fall, creating a cliff.

Sea stacks

3.3 INQUIRY ACTIVITIES

1. Create an annotated diagram that explains the difference between swash and backwash.
   Describing and explaining
2. Use internet resources to find a video or animation on coastal erosion or stack formation. Take note of the process of erosion. **Examining, analysing, interpreting**
3. In small groups, create your own claymation or stop-motion movie, Prezi, or animated PowerPoint to show the *changes* that happen to a cliffed coast eroding to form a notch, cave, arch and stack.
   Classifying, organising, constructing
4. Most Australians live within an hour's drive of the coast, and many people either spend regular holidays on the coast or move to the coast in their retirement, for a 'sea change'. How might the continually *changing* coastal landscape (as seen in **FIGURE 2**) affect coastal housing and popular holiday places? Brainstorm this with a small group. **Evaluating, predicting, proposing**
5. Using a sketch map, identify how several of the *changes* identified in question 4 might affect the coastal landscape of your favourite beach. **Evaluating, predicting, proposing**
6. Rising sea levels, whether they are a naturally occurring process or have resulted from human activity, will affect coastal landscapes. Use a diagram, with annotations, to explain how rising sea levels could *change* two of the landforms illustrated in **FIGURE 2**. **Describing and explaining**

3.3 EXERCISES

Geographical skills key: GS1 Remembering and understanding **GS2** Describing and explaining **GS3** Comparing and contrasting **GS4** Classifying, organising, constructing **GS5** Examining, analysing, interpreting **GS6** Evaluating, predicting, proposing

3.3 Exercise 1: Check your understanding

1. **GS1** What is a coast?
2. **GS1** What are three physical processes that have influenced the creation of coastal landforms?
3. **GS1** What are three human activities that have influenced the creation of coastal landforms?
4. **GS1** Place the following landforms in the order in which they would be created:
   (a) arch, cave, headland, stack
   (b) blowhole, cave, cliff.
5. **GS2** Explain the difference between constructive and destructive waves.

3.3 Exercise 2: Apply your understanding

1. **GS2** Find an image of a sandy coastline that has recently been affected by destructive waves. Explain the process that has occurred. Use the terms *swash* and *backwash* in your explanation.
2. **GS1** What does the construction material that is deposited on a beach consist of?
3. **GS6** Do you think people will still feel the same way about a coastal landscape such as the Twelve Apostles when only two or three are still standing? How might the *changing* landscape affect the value or pleasure people get from visiting this *place*? Write a short paragraph to comment.
4. **GS6** Destructive waves are bad for all coastal *environments* and as such, management techniques should be used to minimise their impacts. Do you agree or disagree with this statement? Justify your response.
5. **GS6** Should we try to protect coastal landforms like the Twelve Apostles or should we simply let nature run its course?

Try these questions in learnON for instant, corrective feedback. Go to www.jacplus.com.au.

3.4 Which coastal landforms are created by deposition?

3.4.1 How are depositional coastal landforms formed?

As we learned in subtopic 3.3, not all waves are destructive. Though they lack the sheer force of destructive waves, constructive waves still have an important role to play in the creation of coastal landforms. The movement of these waves towards the land is more likely to push material such as sand and shells and deposit them on the beach, building new coastal features.

A beach is a good example of a depositional coastal landform (**FIGURE 1**). Sand has been deposited and built up over a period of time. Constructive waves build coastal landscape features by repositioning wave-born materials to also create spits, sand dunes and lagoons.

FIGURE 1 Depositional landforms: coastal landforms created by deposition

A spit can sometimes join two land areas. This is called a tombolo.

Prevailing winds

Longshore drift can build up sand to form a spit.

Inlet

Sea island

Bay mouth bar

Sand bar

Bay

Bay

A lagoon develops when a sandbar closes in an area.

Beach

A beach forms when material is brought to shore by waves. Sand dunes form when plenty of sand builds up on the land.

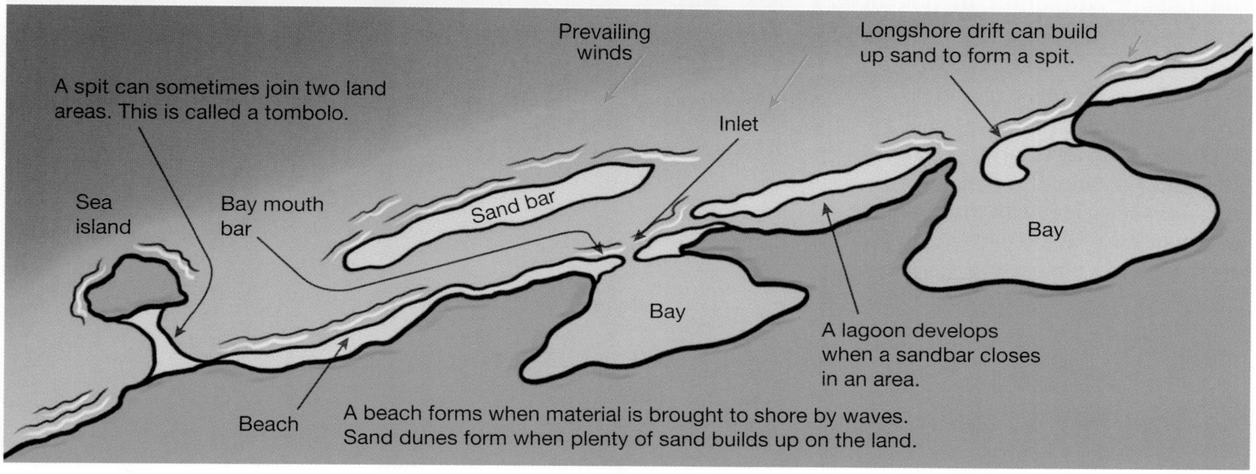

The coastal features created by deposition can be created only when material is brought onshore by the swash of constructive waves. The construction material is in the form of sand, shells, coral and pebbles. The source of the construction material may come from eroding cliffs, from an offshore source, or from rivers which, when they enter the sea, dump any material they were transporting.

This construction material is then shaped by prevailing winds. **FIGURE 2** illustrates the cross-section of a beach formed when there is plenty of sand being pushed onshore by the swash. This construction material is dried by the sun and blown inland to create dunes.

FIGURE 2 The formation of sand dunes

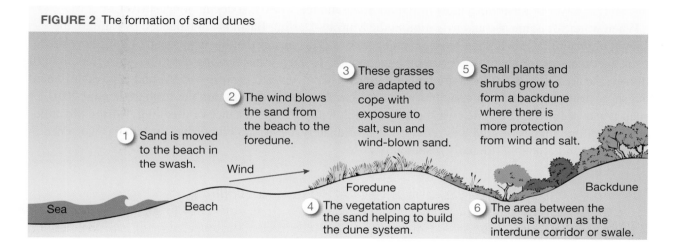

3 These grasses are adapted to cope with exposure to salt, sun and wind-blown sand.

5 Small plants and shrubs grow to form a backdune where there is more protection from wind and salt.

2 The wind blows the sand from the beach to the foredune.

1 Sand is moved to the beach in the swash.

Wind

Sea

Beach

Foredune

4 The vegetation captures the sand helping to build the dune system.

Backdune

6 The area between the dunes is known as the interdune corridor or swale.

Beach material can also be shifted by waves, which get their energy from the wind. The wind influences or directs the angle that waves move towards the coast. Waves come from the direction of the **prevailing wind**. This means that waves often move towards the shore at an angle, and their swash pushes any material they are carrying onto the beach at an angle. As the backwash of the wave returns to the sea, its path takes the shortest possible route down the beach towards the water. This action is known as **longshore drift**, and it is shown in **FIGURE 3**. Longshore drift moves material along the beach in a zigzag pattern that follows the direction of the prevailing wind. Longshore drift moves sand along the beach and creates spits and bars. If the prevailing wind changes direction, then so does the direction of longshore drift.

FIGURE 3 The process of longshore drift

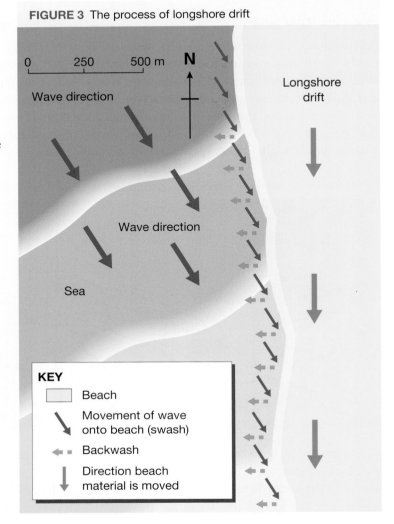

FIGURE 4 Angel Road is a depositional landform connecting three small islands with the mainland in Japan.

3.4.2 CASE STUDY: The Murray mouth, South Australia

The Murray River is Australia's most important river and the world's sixteenth longest river.

When water for home use and irrigation in the Murray–Darling Basin is not balanced by rainfall, the amount of water that reaches the river mouth decreases. This means that the deposition of longshore drift is stronger than the trickle of water reaching the mouth. To keep water flowing out to sea at the mouth of the Murray, the area has undergone a dredging program. This involves removing excess sand from areas where longshore drift has blocked the mouth of the river. The first dredging program ran from 2002 until 2010, when it was deemed that the area was healthy enough without the assistance of dredging. During this time, over 6.5 million tonnes of sand was removed! Unfortunately, the health of the river system has worsened in recent years and dredging officially restarted in 2015.

FIGURE 5 The mouth of the Murray River, South Australia

3.4 INQUIRY ACTIVITIES

1. Draw a sketch of the tombolo shown in **FIGURE 4**.   **Describing and explaining**
2. Use your atlas and the internet to locate and name *places* in Australia that have the following coastal landforms: a spit, a beach with dunes, a bay, a headland (point, cape or promontory) and an estuary. Find four examples of each landform and mark them on a map. You could create a Google map of your results, with links to images of each feature.   **Classifying, organising, constructing**
3. Research water use in the Murray–Darling system and discuss how this may impact the water flow at the mouth of the Murray River.   **Evaluating, predicting, proposing**

3.4 EXERCISES

Geographical skills key: GS1 Remembering and understanding **GS2** Describing and explaining **GS3** Comparing and contrasting **GS4** Classifying, organising, constructing **GS5** Examining, analysing, interpreting **GS6** Evaluating, predicting, proposing

3.4 Exercise 1: Check your understanding

1. **GS1** Where does the material come from that builds beaches?
2. **GS2** The formation of sand dunes cannot happen unless there is plenty of sand in the swash to allow them to grow. Use the information in **FIGURE 2** to provide the evidence for you to agree or disagree with this statement.
3. **GS2** How is weather involved in the formation of sand dune *environments*?
4. **GS1** Describe the process of longshore drift.
5. **GS2** Explain two ways in which the wind can help shape beach *environments*.

3.4 Exercise 2: Apply your understanding

1. **GS5** Study **FIGURE 3**.
   (a) In which direction is sand moving on the beach?
   (b) How will this beach *change* if the longshore drift continues in this direction?
   (c) Redraw this diagram to show how the movement of sand along this beach would *change* this *environment* if the prevailing wind *changed* to come from the south-west.

▶

2. **GS6** Referring to **FIGURE 2**, sketch a new diagram to show what you think would happen to these sand dunes if a fire destroyed the vegetation on the foredune.
3. **GS2** Describe how coastal landforms are the result of *interconnections* between the sea and the atmosphere.
4. **GS6** If it was a windy day, where on the beach or dune would it be best to take shelter? Explain your answer.
5. **GS6** Why did the Murray mouth need to be dredged and do you think this procedure will need to happen again in the future?

Try these questions in learnON for instant, corrective feedback. Go to www.jacplus.com.au.

3.5 Managing coasts

3.5.1 How can a coast be managed?

It is possible to reduce or slow the change to coastal landscapes if we understand the **physical processes** and human activities that cause it. While it is not possible to change the speed and direction of the wind or the number of months each year when destructive waves reach a shoreline, it is possible to redistribute or trap the sand shifted by storm waves or longshore drift. It is also possible to protect coastal houses and roads using barriers to reduce the direct impact of waves.

Coastal management techniques are commonly divided into two main categories — hard engineering strategies and soft engineering strategies. **Hard engineering** strategies typically involve using physical structures to control the effects of natural processes. Sea walls, groynes, gabions and breakwaters are all examples of hard management techniques. What is interesting about these kinds of strategies is that, over time, they can often create problems that are more severe than the ones which they were trying to solve. Let's use a seawall as an example.

Look at **FIGURE 1**. As waves hit the shore in this area, they removed sand from the beach and decreased the stability of the dune system. Concerned that the dunes would eventually be washed away completely, the local council decided to build a sea wall. Although the wall succeeded in protecting the dune, its presence inadvertently caused another management issue. As you can also see in the photograph, there is no sand in front of the sea wall. Before the wall existed, waves did indeed remove sand from the beach and dune system. However, they also replenished the sand over time in a natural cycle. The presence of the wall has interrupted this natural cycle, eventually resulting in the complete loss of beach area in front of the wall. This is just one example of how hard engineering strategies can often cause long-term issues in coastal environments.

FIGURE 1 Sea wall at Brighton Beach

Due to the issues that often arise from hard engineering strategies, many of the strategies we see used today involve **soft engineering** techniques. Taking a more sustainable approach to coastal management, these strategies commonly use natural processes instead of permanent physical infrastructure. Instead of building a sea wall, Bayside Council (responsible for Brighton Beach) could have revegetated the dune system to improve its stability. Dune revegetation is a common soft engineering strategy that involves planting natural grasses and shrubs. As these plants grow, their roots help bind the sand together, halting erosion.

3.5.2 CASE STUDY: Managing Adelaide's living beaches

The problem: The beautiful sandy beaches closest to Adelaide are under constant threat from erosion.
FIGURE 2 identifies the problem. For the past 7000 years the beaches south of Adelaide have been eroding, and the prevailing winds from the south-west have driven this material northwards.

This longshore drift has removed material from the south and relocated it in North Haven, where a **peninsula** has grown and a large dune system has been created. For the past 30 years the beaches in the south have been replenished by adding truckloads of sand. The plan is to find a better way to manage Adelaide's beaches by reducing the cost of moving sand.

The solution: Adelaide's Living Beaches Strategy. **FIGURE 3** illustrates the solution. Although sand will still need to be recycled from north to south, the plan is to use a pipeline instead of trucks to do most of the transportation. The pipeline will extend along the coast and will send sand back to the southern end of the beach. **FIGURE 4** shows sand being discharged at the southern end of the beach. A series of structures such as breakwaters and groynes will be built in several places to trap sand at important locations. Fewer trucks will be used, and it is expected that the cost of beach restoration will be reduced.

FIGURE 2 The movement of sand northwards along the Adelaide Metropolitan coastline

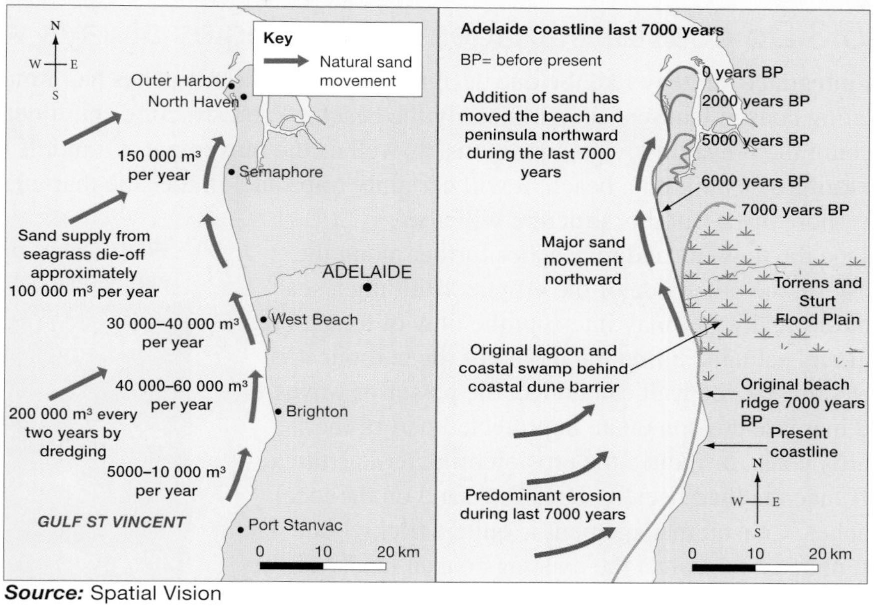

Key
Natural sand movement

Outer Harbor
North Haven

150 000 m³ per year

Semaphore

Sand supply from seagrass die-off approximately 100 000 m³ per year

30 000–40 000 m³ per year

ADELAIDE

West Beach

40 000–60 000 m³ per year

200 000 m³ every two years by dredging

Brighton

5000–10 000 m³ per year

GULF ST VINCENT

Port Stanvac

0   10   20 km

Adelaide coastline last 7000 years

BP= before present

Addition of sand has moved the beach and peninsula northward during the last 7000 years

0 years BP
2000 years BP
5000 years BP
6000 years BP
7000 years BP

Major sand movement northward

Torrens and Sturt Flood Plain

Original lagoon and coastal swamp behind coastal dune barrier

Original beach ridge 7000 years BP

Present coastline

Predominant erosion during last 7000 years

0   10   20 km

Source: Spatial Vision

FIGURE 3 Adelaide's Living Beaches Strategy

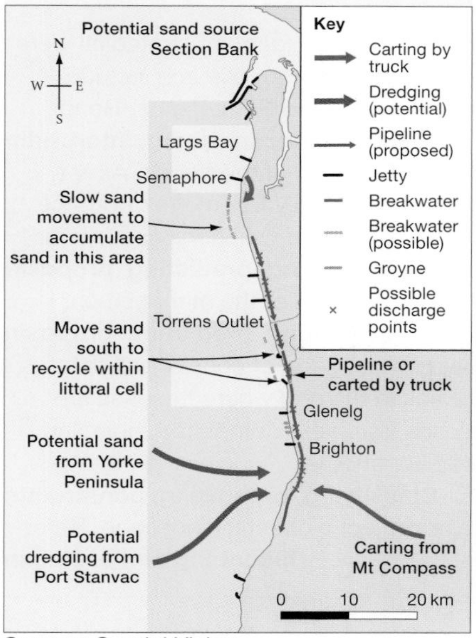

Potential sand source Section Bank

Key
Carting by truck
Dredging (potential)
Pipeline (proposed)
Jetty
Breakwater
Breakwater (possible)
Groyne
Possible discharge points

Largs Bay

Semaphore

Slow sand movement to accumulate sand in this area

Torrens Outlet

Move sand south to recycle within littoral cell

Pipeline or carted by truck

Glenelg

Potential sand from Yorke Peninsula

Brighton

Potential dredging from Port Stanvac

Carting from Mt Compass

0   10   20 km

Source: Spatial Vision

FIGURE 4 Piping sand from north to south along Adelaide's beaches

3.5.3 Do coastal management strategies always work?

An integrated strategy like the one designed for Adelaide's beaches has a much better chance of protecting existing coastal landscapes (particularly the beaches) and structures built nearby, because it has taken into account the prevailing wind conditions, as well as the movement of sand. If a structure like the groyne in **FIGURE 5** is built on a beach, it will certainly trap sand on the side that interrupts the direct flow of the longshore drift. But this structure will also reduce the flow of sand to beaches further along the coast, on the other side of the groyne. Building a sea wall or breakwater may interrupt the flow of longshore drift and actually silt up the mouth of the harbour it is protecting. A sea wall can deflect the power of waves and increase erosion on an unprotected part of the nearby coast, or reduce the erosion of material from a cliff face that had been replenishing sand on the local beaches. Coastal management is quite a tricky issue. Do you manage to protect the existing coastal landscape or do you manage to allow the action of wind and waves to create a naturally evolving landscape?

FIGURE 5 A groyne and rock barrier protect a sandy beach in Wales, United Kingdom.

Groyne

Rock wall

Explore more with myWorldAtlas

Deepen your understanding of this topic with related case studies and questions
• Investigate additional topics > Oceans and coasts > **Managing coasts**

3.5 INQUIRY ACTIVITIES

1. Research another example of coastal landscape management. Identify why the management strategies were put in place and comment on their success. Examples of *places* that would be good to research include Cape Woolamai, the Gold Coast, Melbourne bayside beaches, Polder coastline of the Netherlands, Bondi, Cottesloe, Venice Beach or Waikiki. **Examining, analysing, interpreting**

2. Imagine that you own a holiday house that is built on coastal dunes within 15 metres of the beach. After a powerful storm, the beach in front of your house is eroded and your house is now only five metres from the sea. What are your options? Work out a series of strategies that you could implement which may save your house from falling into the sea. Include diagrams to illustrate your plan. **Evaluating, predicting, proposing**

3. Identify the strengths and weaknesses, for your house and your neighbours' houses, of the management proposal you created to answer question 2. **Evaluating, predicting, proposing**

4. Refer to this and the previous subtopic to make a list of all the uses that can be made of coasts. List these across the top and side of a large table to create a matrix. Now place a tick in the grid where the uses are compatible and a cross where they are not. Choose two incompatible uses from your completed table and work with a partner to develop three criteria that will decide on one use over another. **Classifying, organising, constructing**

5. Draw a diagram to explain how groynes and sea walls help to manage or protect a coastal landscape. Refer to **FIGURE 3** in subtopic 3.4 to help with your diagram. **Describing and explaining**

3.5 EXERCISES

Geographical skills key: GS1 Remembering and understanding **GS2** Describing and explaining **GS3** Comparing and contrasting **GS4** Classifying, organising, constructing **GS5** Examining, analysing, interpreting **GS6** Evaluating, predicting, proposing

3.5 Exercise 1: Check your understanding

1. **GS1** How do groynes and sea walls help to manage or protect a coastal landscape?
2. **GS1** Discuss the two main types of coastal management techniques. Ensure that you explain how they differ.
3. **GS2** What problem do sea walls usually attempt to solve?
4. **GS2** Describe one situation in which you would use a hard management technique instead of a soft management technique.
5. **GS2** Describe one situation in which you would use a soft management technique instead of a hard management technique.

3.5 Exercise 2: Apply your understanding

1. **GS2** Describe what will happen to Adelaide's southern beaches if they stop being replenished with trucks of sand.
2. **GS2** Refer to **FIGURE 2**. Describe the *changes* that have occurred to Adelaide's coastline over the past 7000 years.
3. Refer to **FIGURES 3** and **4**. Describe the *changes* the Living Beaches Strategy has made to the Adelaide coastline and the reasons for these *changes*.
4. **GS2** Draw a diagram to demonstrate how a sea wall is supposed to work.
5. **GS2** Draw a diagram to demonstrate how revegetation could be used instead of a sea wall.

Try these questions in learnON for instant, corrective feedback. Go to www.jacplus.com.au.

3.6 Indigenous use of coastal environments

3.6.1 How did Indigenous Australians use coastal environments?

Indigenous Australians have been using coastal environments for at least 65 000 years. During this time they learned to manage their resources and practised careful and deliberate environmental management techniques. Although the coastal environments we see in Australia today are dramatically different to those used by the first Australians, some archaeological evidence of Indigenous coastal land use does still exist.

Scattered across coastal environments throughout Australia are thousands of fascinating archaeological sites which allow us to examine Indigenous Australian land use. These sites are called **shell middens** and contain the remains of shellfish, bones and sometimes stone tools (see **FIGURE 1**). Shell middens can be found across Australia but are particularly common in New South Wales, Victoria and Tasmania. Shell middens are usually located in scrubland behind sand dunes or in other sheltered positions along a coastline. Aboriginal people used middens to both store and cook their food, as suggested by the presence of heavy amounts of ash and charcoal at these sites. We can use the carbon in these remains to establish the age of individual sites. The oldest Victorian shell midden is located at Cape Bridgewater and was used over 12 000 years ago!

FIGURE 1 Shell midden on the Tarkine coast, Tasmania

While shell middens provide us with important archaeological evidence, they also play an important role in the lives of Indigenous communities today. Physical links to Indigenous heritage are rare and shell middens provide Australian Indigenous peoples with tangible connections to their past. As shell middens are usually situated in delicate and dynamic coastal environments, it is vital that we preserve the historical and cultural significance of these sites.

3.6 INQUIRY ACTIVITY

Shell middens are one example of an Indigenous archaeological site. Use the internet to find another type of Indigenous archaeological site in Australia. Identify and describe the site and explain why it is historically and culturally significant. **Describing and explaining**

3.6 EXERCISES

Geographical skills key: GS1 Remembering and understanding **GS2** Describing and explaining **GS3** Comparing and contrasting **GS4** Classifying, organising, constructing **GS5** Examining, analysing, interpreting **GS6** Evaluating, predicting, proposing

3.6 Exercise 1: Check your understanding

1. **GS1** What is a shell midden?
2. **GS1** Where can Victoria's oldest shell midden be found and how old is it?
3. **GS2** Why are shell middens important to contemporary Indigenous communities?
4. **GS2** In which locations were shell middens found?
5. **GS5** What evidence is there in the middens that suggests Indigenous Australians cooked their food?

3.6 Exercise 2: Apply your understanding

1. **GS2** Most shell middens are found within a few kilometres of a coastline. Why would this location make these sites vulnerable?
2. **GS6** Suggest a way that we could protect and preserve shell middens.
3. **GS6** Suggest how shell middens could be used to boost tourism in regional areas.
4. **GS6** Develop a proposal to the local member for Cape Bridgewater that the shell midden site should be nominated as a location of cultural significance.
5. **GS5** Some middens have been found far from current coastal areas. Suggest how this is possible.

Try these questions in learnON for instant, corrective feedback. Go to www.jacplus.com.au.

3.7 Comparing coastal landforms

3.7.1 How do coastal landforms differ?

Although coastal landforms can be similar in different parts of the world, they can also be very different. Some differences are climatic and some are geomorphic. Coastal landscapes are created by the interconnections between the sculpting power of the oceans, coastal topography and the material that is available to sculpt.

Limestone stacks, such as the Twelve Apostles in Victoria (**FIGURE 1**, subtopic 3.2), have been shaped by the power of the Southern Ocean. Similar stacks have been formed by the erosive power of the waters off the coast of Thailand (**FIGURE 1**) and along the Portuguese and Welsh coasts. We can also compare two regions that feature coastal lake environments — Gippsland Lakes in south-eastern Victoria and the Icelandic Vatnajökull glacier.

FIGURE 1 Ko Tapu rock near Phuket, Thailand

The Gippsland Lakes are a network of coastal lakes and lagoons fed by six rivers but they are often cut off from the sea by a barrier of silt. The Gippsland Lakes are at the mouth of the Mitchell, Avon, Thompson, Latrobe, Nicholson and Tambo Rivers. When there is little rainfall, the rivers flow slowly and deposit sediment in the lakes. This, along with the longshore drifting of the sea current in Bass Strait, creates lakes by moving sediment to seal the lakes with offshore barriers. After heavy rainfall the level of water in the Lakes rises and the barrier breaks, allowing access of fresh water to the sea and salt water into the Lakes. This lake system had an artificial entrance cut by humans in the late 1800s to allow fishing boats into and out of the Gippsland Lakes and to reduce the chance of algal blooms.

In south-eastern Iceland the melting Vatnajökull glacier (**FIGURE 2**) flows into the Atlantic Ocean through a glacial lake.

FIGURE 2 Jökulsárlón Glacier Lagoon, Iceland

This glacier once flowed directly into the sea, but a warming local climate has meant that the glacier's snout is now 1.5 kilometres inland. The melting ice has created the large 18-square-kilometre glacial lake named Jökulsárlón. Since the climate is cold and the sunshine has little heat, the large chunks of ice that fall from the glacier remain as slowly melting icebergs. These icebergs float in the lake until they become small enough to roll down a channel into the sea. During winter the lake freezes and traps the icebergs until the summer thaw. Humans have created a narrow channel to link Jökulsárlón with the sea. This channel is designed to reduce the chance of summer floods and to protect the major highway that brings tourists to this beautiful place.

These two coastal lakes have formed in very different places, with different climates, but the geomorphic process of deposition has meant that human intervention has been required to allow their waters to flow into the sea.

3.7 INQUIRY ACTIVITIES

1. (a) Use the internet to collect at least six images of limestone stacks from different **places** in the world.
   (b) Attach these images to a Google map to create a global distribution of limestone landscapes.
   (c) Describe the similarities and differences between the images. **Comparing and contrasting**
2. Look at a map of the Gippsland Lakes. Predict how they might look if part of the barrier washes away during a huge storm. Draw a sketch map to explain your answer. **Evaluating, predicting, proposing**

3.7 EXERCISES

Geographical skills key: GS1 Remembering and understanding **GS2** Describing and explaining **GS3** Comparing and contrasting **GS4** Classifying, organising, constructing **GS5** Examining, analysing, interpreting **GS6** Evaluating, predicting, proposing

3.7 Exercise 1: Check your understanding

1. **GS1** What material are the Twelve Apostles and Ko Tapu rock both made from?
2. **GS2** How has climate *changed* the entrance of the Vatnajökull glacier into the sea?
3. **GS2** Describe the way that the geological process of deposition has *changed* the Gippsland Lakes and Jökulsárlón.
4. **GS1** How was the Gippsland Lakes area formed?
5. **GS2** What are the key similarities between the Gippsland Lakes and Jökulsárlón?

3.7 Exercise 2: Apply your understanding

1. **GS6** The Vatnajökull glacier is expected to have melted within 80 years. What might this *place* look like when there is no longer a glacier? Draw a sketch map to explain your answer.
2. **GS6** Look at a map of the Gippsland Lakes. Predict how they might look if part of the barrier washes away during a huge storm. Draw a sketch map to explain your answer.
3. **GS5** Explain how rainfall (or the lack of rainfall) can influence the appearance of the Gippsland Lakes region.
4. **GS2** How have humans *changed* the Gippsland Lakes region and Vatnajökull over time?
5. **GS6** What are the major threats to the two regions mentioned in this subtopic? How can these regions be managed to avoid these threats?

Try these questions in learnON for instant, corrective feedback. Go to www.jacplus.com.au.

3.8 How do I undertake coastal fieldwork?

3.8.1 Your fieldwork task

The best way to understand the physical processes and human activities that affect a specific coastal landscape is to visit it. A fieldwork activity will allow you to put the knowledge you have gained in the classroom into practice. Your fieldwork will also allow you to enjoy the coastal landscape in magnificent 3D.

Any coastal landscape would be suitable to investigate. Once a fieldwork site has been identified, there is quite a lot of planning that you should do before you get there.

What is your fieldwork task?

Your task is to identify the landforms and dynamic nature of a coastal landscape and to recognise and assess the influence of people on it.

In class

1. Prepare a base map of the fieldwork site or sites. On this base map, mark in the location of the coastal landscape's natural features (such as beach, rock, dunes, water, vegetation) and **human features** (such as seawall, groyne, steps, lawn, shelter, jetty). Using Google Maps or a topographic map is an excellent way of identifying the specific details of the coastal landscape.
2. Looking at the aerial shot on Google Maps will also allow you to see the pattern of the waves as they move to the shore. Does it look as if longshore drift is occurring on the day this image was taken?

On your field trip

What do you need to do at the coast to collect your information?

It is good to work in groups to collect your data in the field. It is then possible for some students to take measurements and some to record. Sharing tasks means that there will be others with whom to discuss what you have recorded. On returning to class you can pool your observations. You will need recording sheets, pencils, a digital or phone camera, tape measure, compass and maybe a **clinometer**. You could also collect information using data logging equipment, a GPS locator, weather recording equipment and notepads. Your group should decide what equipment is the most practical and relevant for collecting the data you need.

You may not be able to return to your fieldwork site, which means your data needs to be very detailed.

- Always record the location of the information on your map.
- Take photos of the coastal landscape, including the landforms and human structures.
- Measure distances and heights.
- Draw **field sketches** to remind you of details. Even when you have photographed something, a field sketch allows you to annotate the diagram so that you can remember important characteristics about how it was formed or the direction of longshore drift. Do not worry if you are not a gifted artist, as there are apps that allow you to convert your photos to sketches when you get back to class.

FIGURE 1 The information you need to collect at your fieldwork location

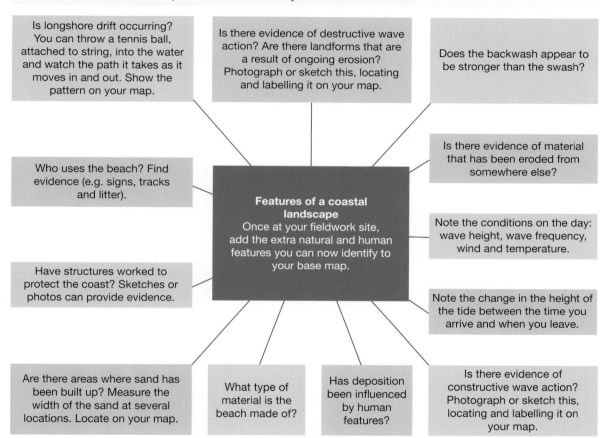

FIGURE 2 Investigating the rocky shores of a coastal landscape

FIGURE 3 Students on a fieldwork trip, measuring the slope of a sandy beach.

Back in class

Now that you have collected your information in the field, you need to present your findings about the coastal landscape you visited.

There are many ways that you could present this information. Your fieldwork report could be presented as a poster, website, PowerPoint presentation, booklet, blog, movie, news report or podcast. Consider using Google Maps and uploading images of the sites you visited. You will need to present the data you collected and describe your findings.

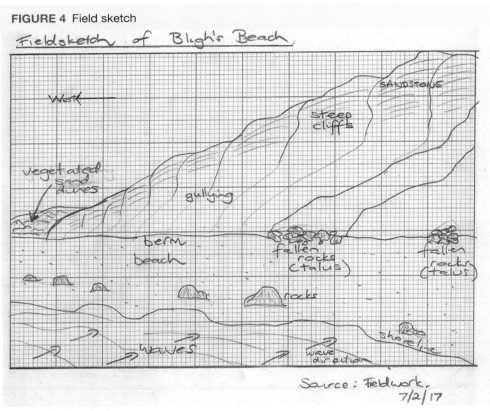

FIGURE 4 Field sketch

Source: Alex Rossimel

3.9 SkillBuilder: Constructing a field sketch

What are field sketches?

Field sketches are drawings completed during fieldwork — geography outside the classroom. Field sketches allow a geographer to capture the main aspects of landscapes in order to edit the view, focusing on the important features and omitting the unnecessary information.

Select your learnON format to access:

- an overview of the skill and its application in Geography (Tell me)
- a video and a step-by-step process to explain the skill (Show me)
- an activity and interactivity for you to practise the skill (Let me do it)
- questions to consolidate your understanding of the skill.

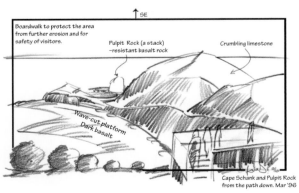

3.10 How does water form river landscapes?

3.10.1 Moving water

Erosion, transportation and deposition are the key processes through which rivers are able to sculpt landscapes. Some rivers, such as the Gordon River in Tasmania, are **perennial**; some, such as Coopers Creek in Queensland, are **intermittent**; others, such as the Colorado River in the United States, have eroded amazing landforms like the Grand Canyon.

Water is always on the move. It evaporates and becomes part of the water cycle; it rains and flows over the surface of the Earth and into streams that make their way to a sea, lake or ocean; and it soaks through the pores of rocks and soil into **groundwater**.

FIGURE 1 Rivers of the world: the longest river on each continent

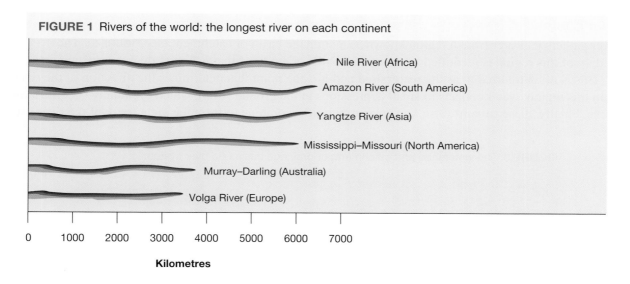

Kilometres

3.10.2 River systems and features

A river is a natural feature, and what we see is the result of the interaction of a range of inputs and processes. All parts of the Earth are related to the formation of river landscapes. This includes the lithosphere (rocks and soil), the hydrosphere (water), the biosphere (plants and animals) and the atmosphere (temperature and water cycle). Changes can happen quickly or over a very long period of time. Changes at one location along a river can have an effect at other locations along the river.

Water flows downhill, and the source (the start) of a river will be at a higher altitude than its mouth (the end). As the water moves over the Earth's surface, it erodes, transports and deposits material.

The volume of water and the speed of flow will influence the amount and type of work carried out by a river. A fast-flowing flooded river will erode enormous amounts of material and transport it **downstream**. As the speed or volume of the water decreases, much of the material it carries will be deposited. Rivers are commonly broken into three main sections – the upper, middle and lower course. Different processes and different types of landforms can be found in each section. Let's examine these sections more closely to see exactly how rivers work.

FIGURE 2 A river system

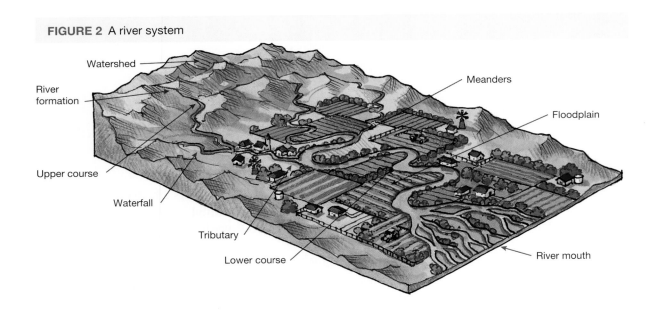

Watershed

River formation

Meanders

Floodplain

Upper course

Waterfall

Tributary

Lower course

River mouth

Upper course

A river gathers its water from a region known as a drainage basin or catchment (see **FIGURE 3**). The boundary of this region is identified by mountains, hills or any land that is higher than the surrounding area. This is often referred to as the **watershed** and it is the point that determines the direction of the river. Within this region, water collects in small depressions in the ground (rills), which eventually become larger streams. Finally, these streams (also known as tributaries) combine to form the main trunk of the river itself.

FIGURE 3 The watershed and catchment, or drainage basin of a river system

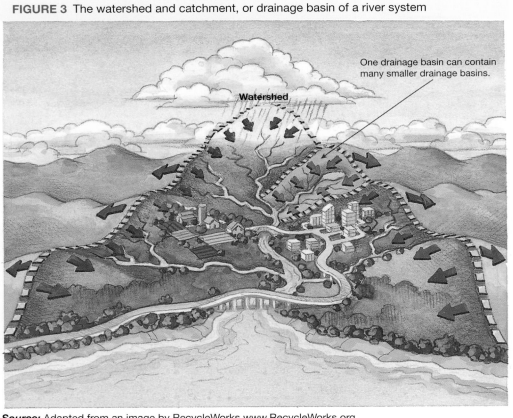

One drainage basin can contain many smaller drainage basins.

Watershed

Source: Adapted from an image by RecycleWorks www.RecycleWorks.org

Water moves quickly along the upper course of a river is it makes its way from areas of higher elevation to areas of lower elevation. The faster the flow of a river, the more power it has and the more erosion it causes. It is common to see waterfalls, plunge pools and rapids along the upper course of a river.

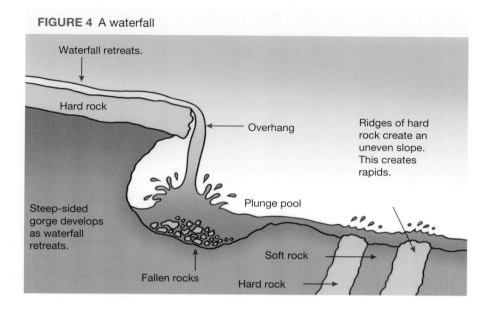

FIGURE 4 A waterfall

Middle course

A river will naturally follow the topography of the surrounding area. As the land flattens out, a river will stretch into long sweeping turns known as **meanders**. Here, the energy of the fast-flowing river we saw in the upper course is converted and allows the river to carve a new path through the flatter landscape of the middle course. Over time, a meandering river will change the path it follows, as some bends become more obvious and others disappear. A meander that has been cut off is called an oxbow lake. In Australia we call these billabongs.

During times of high rainfall, land on either side of the middle course can become inundated as the river struggles to contain excess water. Referred to as a **floodplain**, these areas are highly suitable for agriculture. As floodwaters subside, they leave behind the nutrient-rich sediment (alluvium) that the river had been transporting since it left the upper course.

Lower course

As a river enters the lower course it slows down again, separating back into smaller streams called distributaries. The remaining sediment carried by the river is deposited in an area referred to as the delta. **River deltas** commonly take three main shapes: fan shaped, arrow shaped and bird-foot shaped. The shape of a delta is influenced by tides, waves and the volume of sediment and water carried by a river. Sometimes a river ends with a wide mouth where fresh water and salt water can mix. This is known as an **estuary.**

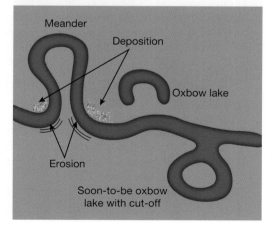

FIGURE 5 The formation of a meander and oxbow lake

DID YOU KNOW
Australia has no major river deltas as a result of the strong ocean currents surrounding the continent.

---

on Resources

🔹 **Interactivity**   River carvings (int-3104)
🔹 **Google Earth**   Mississippi Delta

3.10 INQUIRY ACTIVITIES

1. After some rain, investigate an area of bare ground on a small slope near school or home. Sketch the pattern that the rills have made. Identify the watershed and catchment for each rill.
   Examining, analysing, interpreting
2. Research and then sketch a diagram to show the course of the meandering Murray River. Mark in the course that the river used to take. Predict and label where the next oxbow lake, or billabong, might form. Show the possible future course of the river.   **Evaluating, predicting, proposing**
3. Produce a flowchart or animation to explain the formation of an oxbow lake, a delta, a waterfall or rapids.
   Classifying, organising, constructing
4. Using Google Earth or an atlas, find the Nile delta, the Ebro delta and the Mississippi delta. Draw a sketch and write a short description of the shape of each delta, presenting your findings in a table.
   Classifying, organising, constructing
5. Research river deltas around the world. Discuss any common features between the different areas in which the deltas have formed.   **Comparing and contrasting**

3.10 EXERCISES

Geographical skills key: GS1 Remembering and understanding **GS2** Describing and explaining **GS3** Comparing and contrasting **GS4** Classifying, organising, constructing **GS5** Examining, analysing, interpreting **GS6** Evaluating, predicting, proposing

3.10 Exercise 1: Check your understanding

1. **GS3** Refer to **FIGURE 1** and compare the *scale* of Australia's longest river with the world's longest river.
2. **GS1** What feature, other than water, has to be present for waterfalls and rapids to form? Refer to **FIGURE 4**.
3. **GS2** Explain how rivers are part of the water cycle.
4. **GS2** Why do people settle and farm on floodplains?
5. **GS2** Create a table that explains the positives and negatives of living in a flood plain.

3.10 Exercise 2: Apply your understanding

1. **GS2** Identify a river that flows through the capital city in one state or territory in Australia. Describe its source, any tributaries, and its mouth.
2. **GS6** What do you think will happen to deltas if sea levels rise?
3. **GS6** Predict the *changes* that will occur to the waterfall in **FIGURE 4**. Justify your answer.
4. **GS6** What *changes* will occur along a river if there is unusually high rainfall in its upper course? Think in terms of erosion and deposition.
5. **GS6** Do you think that governments should stop people from living in flood plains? Justify your response.

Try these questions in learnON for instant, corrective feedback. Go to www.jacplus.com.au.

3.11 Managing river landscapes

3.11.1 Mississippi River

Rivers are vital. Plants and animals depend on their waters for survival. People also rely on rivers for their waters and have diverted rivers for flood control, irrigation, power generation, town water supplies, waste disposal and recreation.

The mighty Mississippi River is approximately 3700 kilometres long and is the second longest river in the United States. It flows through 10 states (see **FIGURE 1**). The drainage basin, or catchment, for the river covers 40 per cent of the country, and includes all or part of 31 states and two Canadian provinces. The drainage system is made up of thousands of rivers and streams, including the Missouri.

Importance of the river

The Mississippi has been a major contributor to the economic growth of the United States.

- It is important for transporting goods, such as fuel, coal, gravel, chemicals, steel, cement and farm produce. The **barges** on the river are able to connect to ocean shipping at Baton Rouge in Louisiana.
- It supplies water for cities and industries and irrigation for farming.
- Much of its floodplain has been cleared for farmland.
- The river basin also supports natural biodiversity. It has many species of mussels, 25 per cent of all fish species in North America, and over 300 species of birds that use the river during migration and breeding.

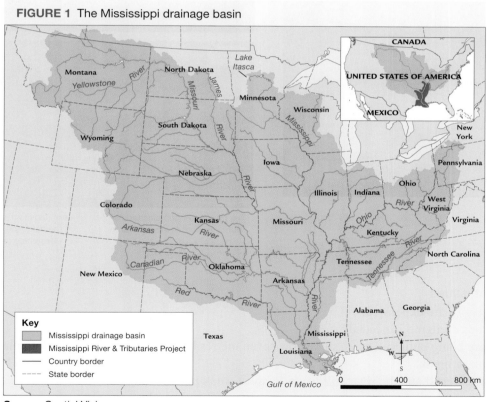

FIGURE 1 The Mississippi drainage basin

Source: Spatial Vision

Floods

The river has created the geographical characteristics that have always attracted settlement. The source of the river is at an altitude of 450 metres above sea level, and the river drops in altitude very quickly. The last 1000 kilometres of the river's journey is through a wide floodplain that is the result of many floods over hundreds of years. Under natural conditions, the river had high water levels in early spring and much lower levels by early autumn.

Floods are a major issue for businesses, homes and farms. There have been many significant floods; for instance, in 1849, 1850, 1882, 1912, 1913, 1927, 1983, 1993 and 2011. After the floods of 1927, the Mississippi River and Tributaries Project was set up with the goal of preventing destructive floods and keeping the river open for navigation.

River management

The Mississippi River and Tributaries Project uses many strategies to manage the river. The aim is to satisfy the needs of farming, towns, industry, transport and **ecosystems**. There are many dams to control water levels in the river.

FIGURE 2 Examples of management strategies

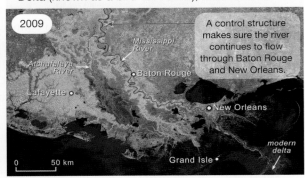

Concreting banks to decrease erosion

A levee bank on each side to build up the height of the natural riverbank

A lock to raise or lower a boat to match the water level on the other side

Dredging to scoop up mud and materials from the riverbed

Straightening of the river to make navigation easier

FIGURE 3 Predicted changes to the Mississippi Delta (known as a bird-foot delta), 2009 to 2100

2009

A control structure makes sure the river continues to flow through Baton Rouge and New Orleans.

Mississippi River

Atchafalaya River

Baton Rouge

Lafayette

New Orleans

0   50 km

Grand Isle

modern delta

2100

Mississippi River

Atchafalaya River

Baton Rouge

Lafayette

New Orleans

0   50 km

Grand Isle

modern delta

The river water contains too many nutrients from towns and farms. This is a concern for the fishing industry.

Less sediment is reaching the delta; the delta is becoming smaller.

Management issues

- The strategies are expensive.
- Continuous dredging is needed.
- Levees are being built higher — some now seven metres high — and it is hard work to make sure they don't leak or break.
- Water is powerful and the river still wears away at weak points along the banks.
- If a levee breaks or if water goes over the top, flood damage can be very bad.
- The floodplain does not receive much sediment from the river.
- The river water is not as clean as it used to be.
- Natural habitats are damaged by dredging or concreting.
- The delta is decreasing in size.

DISCUSS

'Should all buildings be banned from being constructed in a flood plain?' Refer to the issues map and write all the different perspectives that can be included to answer this question. Once complete, categorise these points of view into positive and negative views. **[Critical and Creative Thinking Capability]**

On Resources

🔗 **Weblink** Mississippi

3.11 ACTIVITIES

1. Use the **Mississippi** weblink in the Resources tab to watch a video about the Mississippi River. What do you notice about the *scale* of the watershed and the location of the Mississippi River?

   Examining, analysing, interpreting

2. Do you agree or disagree with the following statement? 'A strategy implemented in one part of the river will have an impact on another part of the river.' As you find evidence from this subtopic, place it in a table, or under subheadings. Write a conclusion based on your findings. **Examining, analysing, interpreting**

3.11 EXERCISES

Geographical skills key: GS1 Remembering and understanding **GS2** Describing and explaining **GS3** Comparing and contrasting **GS4** Classifying, organising, constructing **GS5** Examining, analysing, interpreting **GS6** Evaluating, predicting, proposing

3.11 Exercise 1: Check your understanding

1. **GS2** Refer to **FIGURE 1** and name key tributaries of the Mississippi River. In which general direction does the Mississippi flow from its source to its mouth?
2. **GS4** Why is the river important to the United States? Classify each reason as one or a combination of the following: social, economic or *environmental*.
3. **GS1** How long is the Mississippi River and through how many states does it flow?
4. **GS2** Explain the main two uses of the Mississippi River.
5. **GS1** What are the main issues that engineers face when managing flooding along the Mississippi River?
6. **GS5** Refer to **FIGURE 3.** What does the formation of a bird-foot delta indicate about the type of waves in this part of the Gulf of Mexico?

3.11 Exercise 2: Apply your understanding

1. **GS6** How close will Baton Rouge be to the sea in 2100?
2. **GS6** What do you think would be the main management strategies on the Mississippi River during a year of heavy rainfall? What do you think would be the main management strategies during a drought?
3. **GS2** Explain how the geographic characteristics of the Mississippi River can lead to frequent flooding.
4. **GS6** What kind of human activity occurs the most in the lower course of the Mississippi River? Why does this activity occur in this region?
5. **GS6** What do you believe would be the best flood management strategy (or strategies) to use along the Mississippi River?

Try these questions in learnON for instant, corrective feedback. Go to www.jacplus.com.au.

3.12 Landscapes formed by ice

3.12.1 How can glaciers shape landscapes?

In cold parts of the world, such as the poles and high mountains, water falls as snow, is compacted and then moves more slowly than when it is a liquid. When ice deposits thicken, the same gravitational force that moves flowing water also moves ice, and it begins to flow. Glaciers trace a path downhill from permanent snowfields. The weight of snow and ice crush and scrape surface rocks to produce some distinctive landscapes. Fluctuations in climate cause glaciers to change in length, width and depth, and each change results in alterations to the glacial landscape.

According to the Randolph Glacier Inventory, there are approximately 198 000 glaciers in the world. Predominately found in Antarctica (91 per cent) and Greenland (8 per cent), glaciers make up 0.5 per cent of the Earth's surface (25 million square kilometres). The closest glaciers to Australia are found in the mountains of West Papua (Indonesia) and the alps of New Zealand. There are landscapes caused by glacial activity in Tasmania, although the glaciers themselves have long since disappeared.

FIGURE 1 Franz Josef Glacier in New Zealand's South Island

During the most recent ice age, up to 30 per cent of the Earth's land surface was glaciated. Glaciers have a huge impact on landscapes, and the forces of erosion and deposition they exert are responsible for dramatic changes.

Moraine is any material carried by the glacier. This eroded material may have been picked up from the valley floor or it may have been eroded from the valley wall. Moraine comes in many sizes, from fine silt to very large boulders. As the glacier melts or retreats, it dumps its load of moraine because it no longer has the energy to push it down the slope. **FIGURES 2** and **3** illustrate the movement of the ice as it changes and shapes the environment.

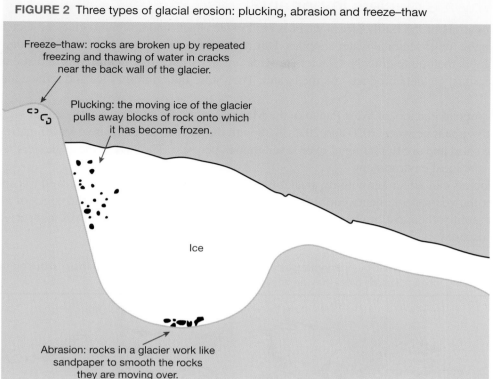

FIGURE 2 Three types of glacial erosion: plucking, abrasion and freeze–thaw

Freeze–thaw: rocks are broken up by repeated freezing and thawing of water in cracks near the back wall of the glacier.

Plucking: the moving ice of the glacier pulls away blocks of rock onto which it has become frozen.

Ice

Abrasion: rocks in a glacier work like sandpaper to smooth the rocks they are moving over.

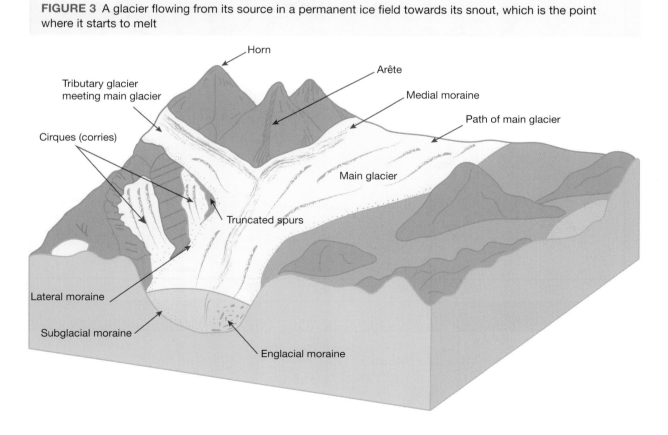

FIGURE 3 A glacier flowing from its source in a permanent ice field towards its snout, which is the point where it starts to melt

Horn

Arête

Medial moraine

Tributary glacier meeting main glacier

Path of main glacier

Cirques (corries)

Main glacier

Truncated spurs

Lateral moraine

Subglacial moraine

Englacial moraine

3.12.2 Why are glaciers important?

About three-quarters of the Earth's fresh water is held in ice sheets and mountain glaciers. Glaciers serve as a natural regulator of regional water supplies. During periods of warm weather, or during dry seasons or droughts, glaciers melt quite quickly. Glaciers provide a water source that feeds rivers and streams. During cold, rainy seasons, glaciers produce less meltwater. They store the rainfall as ice and reduce the chance of a **flash flood**.

The small tropical glaciers of West Papua in the Maoke Mountains of the western central highlands are predicted to disappear between 2020 and 2025. Although these small glaciers are over 30 metres deep, they are quite short and are retreating at over seven metres per year. The loss of these glaciers will result in changes to the local environment.

Melting glaciers can affect agriculture, availability of fresh water, hydroelectric power, transportation and tourism. Over the years, settlements, farming and tourism have extended towards the edges of glaciers. If glaciers melt rapidly, then **avalanches** and flash floods will increasingly threaten lives and services in high mountain landscapes.

FIGURE 4 The state of the world's continental glaciers, not including polar glaciers. If their colour on the map is blue they are losing ice; if it is dark blue, they are losing a lot.

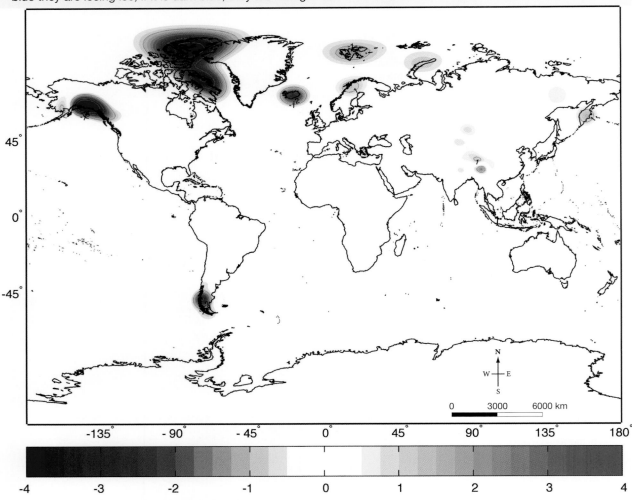

Source: © NASA/JPL-Caltech/University of Colorado

FIGURE 5 Gangotri Glacier in the Himalayas in northern India

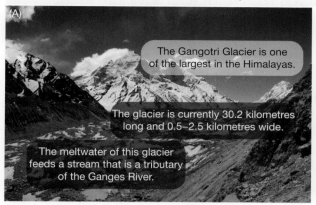

(A)
The Gangotri Glacier is one of the largest in the Himalayas.

The glacier is currently 30.2 kilometres long and 0.5–2.5 kilometres wide.

The meltwater of this glacier feeds a stream that is a tributary of the Ganges River.

(B)
Gangotri Glacier has been receding since 1780, although studies show its retreat has increased since 1971. It has retreated 850 metres in the last 25 years.

Source: © NASA image by Jesse Allen, Earth Observatory; based on data provided by the ASTER Science Team. Glacier retreat boundaries courtesy the Land Processes Distributed Active Archive Center.

  Resources

🔗 **Weblink** Glacier

3.12 INQUIRY ACTIVITIES

1. Complete some internet research to discover how the polar glaciers of Greenland and Antarctica are *changing*. **Evaluating, predicting, proposing**
2. Use the **Glacier** weblink in the Resources tab, and watch the interactivity. Describe the changes that occurred to the glacier over the seven-year period. **Describing and explaining**
3. With reference to the text and images within this subtopic, sketch a diagram or find a suitable image online of a glacier with at least one tributary and annotate the following features: terminal, medial and lateral moraines, arête, cirque, high mountain peaks, glacial stream, U-shaped valley. **Describing and explaining**

3.12 EXERCISES

Geographical skills key: GS1 Remembering and understanding **GS2** Describing and explaining **GS3** Comparing and contrasting **GS4** Classifying, organising, constructing **GS5** Examining, analysing, interpreting **GS6** Evaluating, predicting, proposing

3.12 Exercise 1: Check your understanding

1. **GS1** What is the difference between *plucking, abrasion* and *freeze–thaw*?
2. **GS1** Why do glaciers move?
3. **GS2** Refer to **FIGURE 4**.
   (a) Describe the *places* where glaciers are retreating.
   (b) Describe the *places* where glaciers are advancing.
4. **GS2** Describe one major impact of increased glacial melting.
5. **GS2** How do we get most of our evidence to determine that glaciers are decreasing in size?

3.12 Exercise 2: Apply your understanding

1. **GS6** Check the location of the West Papuan glaciers in your atlas. What is surprising about the *place* these glaciers are found? *Hint:* Look at the latitude.
2. **GS6** How might the local landscape *change* if the glaciers of West Papua melt? How will this *environment change* affect the local inhabitants?
3. **GS6** What can be done to prevent increased glacial melting?
4. **GS2** Describe two reasons why glaciers are important for human populations.
5. **GS5** According to **FIGURE 4**, which areas of the world are losing the most glaciers?

Try these questions in learnON for instant, corrective feedback. Go to www.jacplus.com.au.

3.13 SkillBuilder: Reading contour lines on a map

What are contour lines?

Contour lines drawn on the map join all places of the same elevation (height) above sea level. Contour maps are used to show the relief (shape) of the land and the heights of the landscape. Maps with contour lines show the relief of the land and help people to identify features.

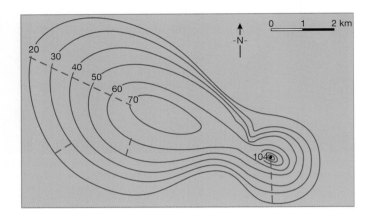

Select your learnON format to access:

- an overview of the skill and its application in Geography (Tell me)
- a video and a step-by-step process to explain the skill (Show me)
- an activity and interactivity for you to practise the skill (Let me do it)
- questions to consolidate your understanding of the skill.

on Resources

Video eLesson SkillBuilder: Reading contour lines on a map (eles-1651)

Interactivity SkillBuilder: Reading contour lines on a map (int-3147)

3.14 Thinking Big research project: Coastal erosion animation

SCENARIO

Unless you are lucky enough to be watching at the *exact* moment that a sea-stack tumbles into the ocean, it can be difficult to catch erosion in action. In this task, you will do what few people before you have achieved — you will capture the impacts of erosion on film by creating an animation that shows how a coastal landform is created.

Select your learnON format to access:

- the full project scenario
- details of the project task
- resources to guide your project work
- an assessment rubric.

on Resources

 projectsPLUS Thinking Big research project: Coastal erosion animation (pro-0169)

3.15 Review

3.15.1 Key knowledge summary
Use this dot point summary to review the content covered in this topic.

3.15.2 Reflection
Reflect on your learning using the activities and resources provided.

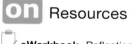 Resources

✓ **eWorkbook**   Reflection (doc-31346)

Crossword (doc-31347)

Interactivity   Landscapes formed by water crossword (int-7596)

KEY TERMS

avalanche a sudden downhill movement of material, especially snow and ice

backwash the movement of water from a broken wave as it runs down a beach returning to the ocean

barge a long flat-bottomed boat used for transporting goods

clinometer an instrument used for measuring the angle or elevation of slopes

deposition the laying down of material carried by rivers, wind, ice and ocean currents or waves

destructive wave a large powerful storm wave that has a strong backwash

downstream nearer the mouth of a river, or going in the same direction as the current

ecosystem an interconnected community of plants, animals and other organisms that depend on each other and on the non-living things in their environment

erosion the wearing away and removal of soil and rock by natural elements, such as wind and water, and by human activity

estuary the wide part of a river at the place where it joins the sea

field sketch a diagram with geographical features labelled or annotated

flash flood a flood that occurs very quickly, often without advance warning

floodplain an area of low-lying ground adjacent to a river, formed mainly of river sediments and subject to flooding

groundwater water that seeps into soil and gaps in rocks

hard engineering a coastal management technique that involves using physical structures to control the effects of natural processes

human features structures built by people

intermittent describes a stream that does not always flow

longshore drift a process by which material is moved along a beach in the same direction as the prevailing wind

meander a winding curve or bend in a river

moraine rocks of all shapes and sizes carried by a glacier

peninsula land jutting out into the sea

perennial describes a stream that flows all year

physical process continuing and naturally occurring actions such as wind and rain

prevailing wind the main direction from which the wind blows

river delta a landform created by deposition of sediment that is carried by a river as the flow leaves its mouth and enters slower-moving or stagnant water. Can take three main shapes: fan shaped, arrow shaped and bird-foot shaped.

shell middens Indigenous archaeological sites where the debris associated with eating shellfish and similar foods has accumulated over time

soft engineering a coastal management technique where the natural environment is used to help reduce coastal erosion and river flooding

swash the movement of water in a wave as it breaks onto a beach

tributary a river or stream that flows into a larger river or lake

watershed an area or ridge of land that separates waters flowing to different rivers, basins or seas

---

4 Desert landscapes

4.1 Overview

Hot and sandy? Cold and windy? What are the features of a landscape that make it a desert?

4.1.1 Introduction

Approximately one-third of the Earth's land surface is desert — arid land with little rainfall. These arid regions may be hot or cold. The actions of wind and, sometimes, water shape the rich variety of landscapes found there. Deserts can be very inhospitable places where conditions make it difficult for people to survive in them. Yet there are many desert locations in which people can and do live. In this topic we will learn about different types of deserts, how they form, their locations around the world, and how people use them.

 **Resources**

☑ **eWorkbook**   Customisable worksheets for this topic

▣ **Video eLesson** Desertscapes (eles-1625)

LEARNING SEQUENCE

  4.1 Overview
  4.2 What is a desert?
  4.3 **SkillBuilder:** Using latitude and longitude   `online only`
  4.4 How the climate forms deserts
  4.5 The processes that shape desert landforms
  4.6 Characteristics of Australia's deserts
  4.7 **SkillBuilder:** Calculating distance using scale   `online only`
  4.8 How did Lake Mungo become dry?
  4.9 How people use deserts
4.10 Antarctica — a cold desert
4.11 **Thinking Big research project:** Desert travel brochure   `online only`
4.12 **Review**   `online only`

To access a pre-test and starter questions and receive immediate, **corrective feedback** and **sample responses** to every question, select your learnON format at www.jacplus.com.au.

4.2 What is a desert?

4.2.1 Defining a desert

A desert is a hot or cold region with little or no rainfall. Around one-third of the Earth's surface is desert and is home to about 300 million people.

Although they receive little rainfall, most deserts receive some form of precipitation. When it does rain, it is usually during a few heavy storms that last a short time.

Hot deserts

Most of the world's hot deserts are located between the Tropic of Cancer and the Tropic of Capricorn (see **FIGURE 3**). They have very hot summers and warm winters. Temperature extremes are common, because cloud cover is rare and **humidity** is very low; this means there is nothing to block the heat of the sun during the day, or prevent its loss at night. Temperatures can range between around 45 °C and –15 °C in a 24-hour period.

TABLE 1 Types of deserts

Rainfall (mm/year)	Type of desert	Examples
< 25	Hyper-arid	Namib; Arabian
25–200	Arid	Mojave
200–500	Semi-arid	Parts of Sonoran Desert

FIGURE 1 The Sahara, an example of a hot desert

Cold deserts

Cold deserts lie on high ground generally north of the Tropic of Cancer and south of the Tropic of Capricorn (see **FIGURE 3**). They include the polar deserts. Any precipitation falls as snow. Winters are very cold and often windy; summers are dry and cool to mild.

FIGURE 2 The Gobi, an example of a cold desert

4.2.2 Deserts of the world

FIGURE 3 The distribution of the world's deserts

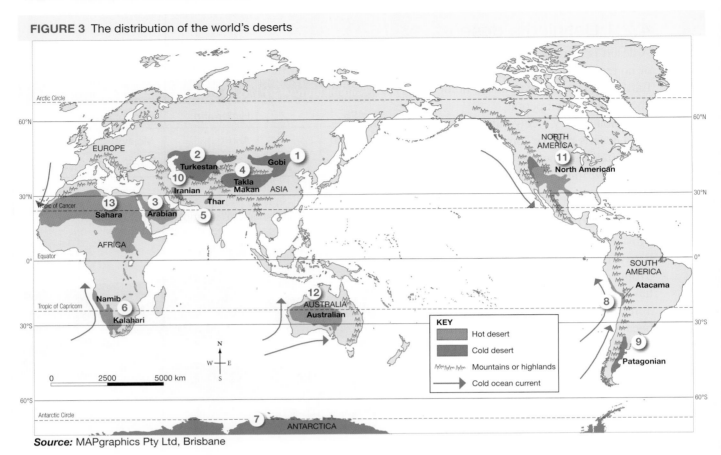

Source: MAPgraphics Pty Ltd, Brisbane

1. **Gobi Desert:** Asia's biggest desert, the Gobi, is a cold desert. It sits some 900 metres above sea level and covers an area of some 1.2 million square kilometres. Its winters can be freezing.

2. **Turkestan Desert:** The cold Turkestan Desert covers parts of south-western Russia and the Middle East.

3. **Arabian Desert:** This hot desert is as big as the deserts of Australia. Towards its south is a place called Rub al-Khali (meaning 'empty quarter'), which has the largest area of unbroken sand dunes, or erg, in the world.

4. **Takla Makan Desert:** The Takla Makan Desert is a cold desert in western China. Its name means 'place of no return'. The explorer Marco Polo crossed it some 800 years ago.

5. **Thar Desert:** The Thar Desert is a hot desert covering north-western parts of India and Pakistan. Small villages of around 20 houses dot the landscape.

6. **Kalahari and Namib deserts:** The Namib Desert extends for 1200 kilometres down the coast of Angola, Namibia and South Africa. It seldom rains there, but an early-morning fog often streams across the desert from the ocean. The dew it leaves behind provides moisture for plants and animals. It joins the Kalahari Desert, which is about 1200 metres above sea level.

7. **Antarctic Desert:** The world's biggest and driest desert, the continent of Antarctica, is another cold desert. Only snow falls there, equal to about 50 millimetres of rain per year.

8. **Atacama Desert:** The Atacama Desert is the driest hot desert in the world. Its annual average rainfall is a tiny 0.1 millimetre.

9. **Patagonian Desert:** The summer temperature of this cold desert rarely rises above 12 °C. In winter, it is likely to be well below zero, with freezing winds and snowfalls.

10. **Iranian Desert:** Two large deserts extend over much of central Iran. The Dasht-i-Lut is covered with sand and rock, and the Dasht-i-Kavir, mainly in salt. Both have virtually no human populations.

11. **North American deserts:** The desert region in North America is made up of the Mojave, Sonoran and Chihuahuan deserts (all hot deserts) and the Great Basin (a cold desert). The Great Basin's deepest depression, Death Valley, is the lowest point in North America.

12 **Australian deserts:** After Antarctica, Australia is the driest continent in the world. Its deserts are generally flat lands, often vibrant in colour.

13 **Sahara Desert:** The largest hot desert in the world, the Sahara stretches some nine million square kilometres across northern Africa over 12 countries. Only a small part is sandy. It is the sunniest place in the world.

On Resources

Interactivity Great deserts of the world (int-3106)

4.2 INQUIRY ACTIVITIES

1. Use the information in this subtopic to design a quiz of 10 questions entitled 'Deserts of the world'. Test your friends and family.   **Classifying, organising, constructing**

2. Draw up and complete a table like the one below to show your understanding of the locations and features of desert **environments**. Look for photos on the internet.   **Classifying, organising, constructing**

Name of desert	Mountain range	Continent	Ocean current	Photos

4.2 EXERCISES

Geographical skills key: GS1 Remembering and understanding **GS2** Describing and explaining **GS3** Comparing and contrasting **GS4** Classifying, organising, constructing **GS5** Examining, analysing, interpreting **GS6** Evaluating, predicting, proposing

4.2 Exercise 1: Check your understanding

1. **GS1** What climate conditions are needed for hot and cold deserts to form?
2. **GS1** Where is the sunniest **place** in the world?
3. **GS1** Name three deserts in the Asia–Pacific region.
4. **GS2** Describe key differences between hot and cold deserts.
5. **GS1** On what major line of latitude are Australian deserts located?

4.2 Exercise 2: Apply your understanding

1. **GS5** Look carefully at the map in **FIGURE 3** and read the text.
   (a) Which continent has the largest area of hot desert?
   (b) Which continent has the largest area of cold desert?
2. **GS5** Look carefully at the map in **FIGURE 3** and read the text.
   (a) What is the largest hot desert in the world?
   (b) What is the largest hot desert in the Asia–Pacific region?
3. **GS5** Look carefully at the map in **FIGURE 3** and read the text.
   (a) Which is the driest continent in the world?
   (b) Which continent contains the driest hot desert?
4. **GS5** Look carefully at the map in **FIGURE 3** and read the text. Which North American desert contains the lowest land on the continent?
5. **GS1** Name the three deserts in Africa and where they are located.

Try these questions in learnON for instant, corrective feedback. Go to www.jacplus.com.au.

4.3 SkillBuilder: Using latitude and longitude

What is latitude and longitude?

Latitude and longitude are imaginary grid lines encircling the Earth. The lines that run parallel to the equator are called parallels of latitude and are measured in degrees. Lines of longitude run from north to south from the North Pole to the South Pole. These are called meridians of longitude and are also measured in degrees. Lines of latitude and longitude are drawn on maps to help us locate places.

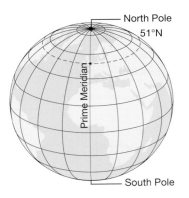

Select your learnON format to access:

- an overview of the skill and its application in Geography (Tell me)
- a video and a step-by-step process to explain the skill (Show me)
- an activity and interactivity for you to practise the skill (Let me do it)
- questions to consolidate your understanding of the skill.

 Resources

Video eLesson SkillBuilder: Using latitude and longitude (eles-1652)

Interactivity SkillBuilder: Using latitude and longitude (int-3148)

4.4 How the climate forms deserts

4.4.1 The subtropics

Deserts form in many different parts of the globe: the subtropics; continental interior areas at middle latitudes; on the leeward side of mountain ranges; along coastal areas; and in the polar regions. The only common factor is their low rainfall — but why do these areas experience low rainfall?

Most of the world's greatest deserts are found in the subtropics near the Tropics of Cancer and Capricorn.

Because of the way the Earth rotates around the sun, areas around the equator receive more direct sunlight than anywhere else on Earth. This means the air there is always very hot. Hot air can hold much more moisture than cold air, so the humidity in these areas is always very high. (If you have ever visited or live near a tropical rainforest or northern Australia, you will have experienced this hot humidity.) Hot air also rises. As the air heads upwards into the atmosphere above the equator, it drifts away, heading north and south.

FIGURE 1 The formation of subtropical deserts in Africa

Warm air rises above equator and cools. Rainforest occurs.

Dry, warm air moves towards tropics.

cools

Dry, warm air moves towards tropics.

Desert

Tropic of Cancer

Tropical forest

warm air

Desert

Equator

Tropic of Capricorn

The higher the air gets, the cooler it becomes. Cool air can't hold as much moisture, so it releases it as rain. Areas around the equator and to the immediate north and south of it (the tropics) receive frequent heavy downpours (see **FIGURE 1**).

With its moisture gone, the cool, dry air continues moving north and south away from the equator until it meets zones of high air pressure around the tropics. Here, it is forced downwards. The more the dry air descends, the warmer it gets. This means it can hold more moisture and it is likely to absorb any moisture that already exists in this environment. It is like using a sponge to wipe up some water on the kitchen bench; a dry sponge will absorb more of the spill than a wet sponge. This is how the subtropical deserts form.

Temperatures in these deserts are usually high all year round. In summer the heat is extreme, with daytime temperatures often going above 38 °C and sometimes as high as 49 °C. At night — with no clouds to provide insulation — temperatures drop quickly to an average of 21 °C in summer and sometimes below freezing during winter.

FIGURE 2 The Sahara Desert of northern Africa is the world's largest and can experience temperatures as high as 57 °C

4.4.2 Rain-shadow deserts

Rain shadows form on the leeward side of a mountain range (opposite the windward side that faces rain-bearing winds). Deserts commonly form in rain shadows.

- Moist air blowing in from the ocean is forced to rise up when it hits a range of mountains. This cools it down. As cool air cannot hold as much moisture, it releases it as precipitation (see **FIGURE 3**).

FIGURE 3 The formation of rain-shadow deserts

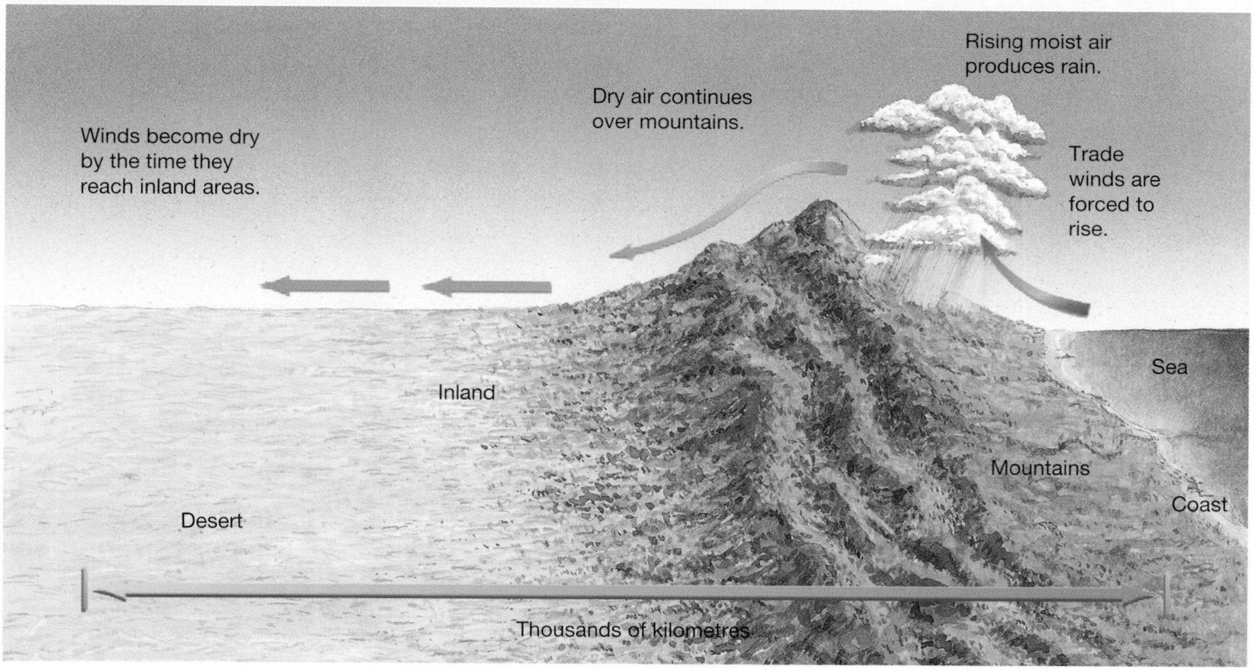

Rising moist air produces rain.

Dry air continues over mountains.

Winds become dry by the time they reach inland areas.

Trade winds are forced to rise.

Inland

Sea

Mountains

Desert

Coast

Thousands of kilometres

- By the time the air moves over the top of the range and down the other side, it is likely to have lost most, if not all, of its moisture. It will therefore be fairly dry.
- The more the air descends on the other side of the range, the more it warms up. Hence, it can hold more moisture. So, as well as not bringing any rain to the land, the air absorbs what little moisture the land contains.
- In time, as this pattern continues, the country in the rain shadow of the mountain range is likely to become arid.

An example of this is the Great Dividing Range in Australia; cool moist air produces winds on the eastern side of these mountains and desert to the west. The Mojave Desert in the south-western United States is located on the leeward side of the Sierra Nevada mountain range (**FIGURE 4**).

FIGURE 4 The Mojave Desert, United States

4.4.3 Coastal deserts

Currents in the oceans are both warm and cold, and are always moving. Cold currents begin in polar and temperate waters (with moderate temperatures), and drift towards the equator. They flow in a clockwise pattern in the northern hemisphere, and in an anticlockwise pattern in the southern hemisphere. As they move, they cool the air above them (see **FIGURE 5**).

FIGURE 5 The formation of coastal deserts

Fog or rain occurs at sea, or on the coast.

Dry air blows inland.

Cooler air

Cold ocean current

If cold currents flow close to a coast, they can contribute to the creation of a desert. This occurs because cold ocean currents cause the air over the coast to become stable, which stops cloud formation. If the cool air the currents create blows in over warm land, the air warms up; it can then hold more moisture. It is therefore not likely to release any moisture it contains unless it is forced up by a mountain range. Large coastal deserts, including the Atacama Desert in Chile (**FIGURE 6**) and the Namib Desert in Namibia (**FIGURE 7**), are formed in this way. The Atacama Desert is a coastal desert in northern Chile in South America and is the driest desert in the world. It is located on the leeward side of the Chilean Coast Range. In some areas, only around one millimetre of rain falls every 5–20 years.

FIGURE 6 The Atacama Desert in Chile, South America

FIGURE 7 The coastal Namib Desert

4.4.4 Inland deserts

Some deserts form because they are so far inland that they are beyond the range of any rainfall. By the time winds reach these dry centres, they have dumped any rain they were carrying or have become so warm they cannot release any moisture they still hold. The air that enters such areas is usually extremely dry and the skies are cloudless for most of the year. Summer daytime temperatures can rise as high as those of subtropical deserts. In winter, however, temperatures are much lower. Average daily temperatures below freezing are common during winter.

Examples of inland deserts are the central deserts of Australia (see **FIGURE 8**), the Thar Desert in north-west India and the vast Gobi and Takla Makan deserts of Central Asia.

FIGURE 8 The Simpson Desert in central Australia

4.4.5 Polar deserts

Polar deserts are areas with a precipitation rate of less than 250 millimetres per year and an average temperature lower than 10 °C during the warmest month of the year. Polar deserts cover almost five million square kilometres of our planet and consist mostly of rock or gravel plains. Snow dunes may be present in areas where precipitation occurs. Temperatures in polar deserts often alternate between freezing and thawing, a process that can create patterned textures on the ground as much as five metres across.

FIGURE 9 Although covered in frozen water, Antarctica receives little rain and is therefore classified as a desert.

DISCUSS
Climate change is already leading to increasing areas of desertification. How important is it for Australians to consider the impact of their high carbon-producing lifestyle on the impact of such landscapes?

[Critical and Creative Thinking Capability]

4.4.6 Desert climate

Temperature

One geographical characteristic of many deserts is the high temperature, which quickly evaporates any water that might be around. The Earth's highest recorded temperature — 56.7 °C — occurred at Greenland Ranch in Death Valley, California, United States on 10 July 1913.

During the summer of 1923–24, the semi-arid town of Marble Bar in Western Australia (average rainfall 361 mm per year) experienced temperatures of more than 37.8 °C for 160 days in a row, from 31 October 1923 to 7 April 1924. However, the highest official maximum temperature recorded in Australia was 50.7 °C at Oodnadatta in South Australia on 2 January 1960.

Rainfall

Although low rainfall is a characteristic of deserts, rain does fall and violent storms can sometimes occur. A record 44 millimetres of rain once fell within three hours in the Sahara. Large Saharan storms may deliver up to one millimetre of rain per minute. Normally dry stream channels, called arroyos or wadis, can quickly fill after heavy rains, and flash floods make these channels dangerous.

Monthly data for rainfall and temperature can be used to create climographs for other desert locations such as Khormaksar in Yemen and Alice Springs in Australia (see **TABLE 1**).

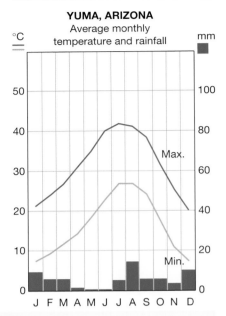

FIGURE 10 Yuma, Arizona climograph

TABLE 1 Climate data for (a) Khormaksar, Yemen, and (b) Alice Springs, Australia

(a) Khormaksar, Yemen

	Jan	Feb	Mar	Apr	May	Jun	Jul	Aug	Sep	Oct	Nov	Dec	Total
Average temperature (°C)	25.0	25.5	27.0	28.5	30.5	33.0	32.0	32.0	32.0	28.5	26.5	25.5	
Average rainfall (mm)	5.0	0.0	5.0	0.0	0.0	0.0	5.0	3.0	0.0	0.0	0.0	5.0	23.0

(b) Alice Springs, Australia

	Jan	Feb	Mar	Apr	May	Jun	Jul	Aug	Sep	Oct	Nov	Dec	Total
Average temperature (°C)	28.5	27.7	24.8	20.0	15.4	12.4	11.5	14.3	18.3	22.8	25.8	27.7	
Average rainfall (mm)	40.5	41.5	34.7	16.6	17.0	16.7	12.1	10.0	9.0	20.0	25.3	37.2	280.6

On Resources

Interactivity How to make a desert (int-3107)

Weblink Desert rain

Google Earth Alice Springs
Yemen

4.4 INQUIRY ACTIVITY

Use the **Desert rain** weblink in the Resources tab, and then answer the following questions.
a. What is a flash flood?
b. What happens to water as it flows over sand? Think of what happens to water at the beach.

c. How do animals and plants respond to these rare water events?

d. Describe how the landscape quickly *changes* once there is water in the desert. **Describing and explaining**

4.4 EXERCISES

Geographical skills key: GS1 Remembering and understanding **GS2** Describing and explaining **GS3** Comparing and contrasting **GS4** Classifying, organising, constructing **GS5** Examining, analysing, interpreting **GS6** Evaluating, predicting, proposing

4.4 Exercise 1: Check your understanding

1. **GS1** Decide whether the following statements are true or false. Rewrite the false statements to make them true.
   (a) The cooler the air, the more moisture it can hold.
   (b) Rain shadows often contain dry areas of land.
   (c) Cold ocean currents cool the air above them.
   (d) Deserts do not form along coastlines.
2. **GS2** Use **FIGURE 1** to explain why deserts form around areas near the tropics but not at the equator. Alternatively, form small groups and create a short drama performance to explain the process.
3. **GS2** Use **FIGURE 3** and any other information in this subtopic to write a paragraph explaining why deserts tend to form in rain shadows. Alternatively, form small groups and create a short drama performance to explain the process.
4. **GS1** Why do temperatures in deserts drop so much at night after being so high during the day?
5. **GS1** What are the extremes of temperatures that have been recorded in hot deserts?

4.4 Exercise 2: Apply your understanding

1. **GS2** Draw a diagram to explain how cold ocean currents influence the formation of a desert *environment* along the Western Australian coastline.
2. **GS4** Use **TABLES 1a** and **1b** to draw climate graphs for Khormaksar, Yemen, and Alice Springs, Australia.
3. **GS1** What are the characteristics of polar deserts?
4. **GS1** Give four examples of inland deserts.
5. **GS2** Describe how inland deserts form.

Try these questions in learnON for instant, corrective feedback. Go to www.jacplus.com.au.

4.5 The processes that shape desert landforms

4.5.1 Shaping the desert

Although most people imagine a sea of sand when they think of deserts, sand covers only about 20 per cent of the world's deserts. Sand is the end product of millions of years of erosion of other landforms such as rock and plateaus that, over time, are worn away by extremes of temperature, wind and water.

The landforms and patterns of a desert are created by a number of natural processes. The unprotected land surfaces are prone to erosion. After heavy rain, often a long distance from the desert flood plains, erosion of ancient river channels can be major. Extreme temperatures, along with strong winds and the rushing water that can follow a desert rainstorm, cause rocks to crack and break down into smaller fragments. This process is called weathering.

Erosional landforms

The process of erosion removes material such as weathered rock. Most erosion in deserts is caused by wind and, at times, running water. During heavy rainfall, water carves channels in the ground. Fast-flowing water can carry rocks and sand, which help to scour the sides of the channel. As vegetation is usually sparse or non-existent, there are few roots to hold the soil together. Eventually, deep gullies called wadis can form.

Erosion can also result from the action of wind and from chemical reactions. Some rock types, such as limestone, contain compounds that react with rainwater and then dissolve in it. Wind is a very important agent of transport and deposition, and can change the shape of land by abrasion — the wearing down of surfaces by the grinding and sandblasting action of windborne particles.

Depositional landforms

Materials carried along by rushing water and wind must eventually be put down. Over time these materials build up, forming different shapes and patterns in the desert. This process is called deposition.

Depositional landforms in deserts include alluvial fans, playas, saltpans and various types of sand dunes (see **FIGURE 1**).

FIGURE 1 Desert landforms

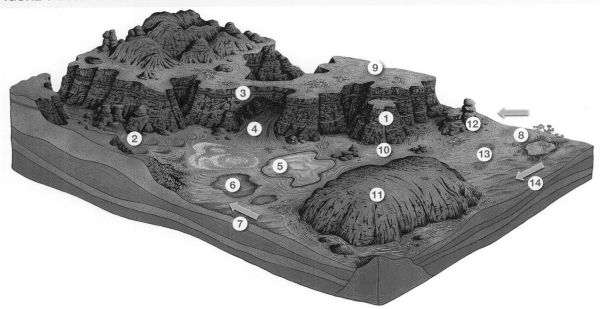

1. A butte is the remaining solid core of what was once a mesa. It often is shaped like a castle or a tower.
2. Crescent-shaped barchan dunes are produced when sand cover is fairly light.
3. An arch, or window, is an opening in a rocky wall that has been carved out over millions of years by erosion.
4. An alluvial fan is the semicircular build-up of material that collects at the base of slopes and at the end of wadis after being deposited there by water and wind.
5. A playa lake may cover a wide area, but it is never deep. Most water in it evaporates, leaving a layer of salt on the surface. These salt-covered stretches are called saltpans.
6. Clay pans are low-lying sections of ground that may remain wet and muddy for some time.
7. The rippled surface on transverse dunes is the result of a gentle breeze blowing in the one direction.
8. An oasis is a fertile spot in a desert. It receives water from underground supplies.
9. A mesa is a plateau-like section of higher land with a flat top and steep sides. The flat surface was once the ground level, before weathering and erosion took their toll.
10. Sand dunes often start as small mounds of sand that collect around an object such as a rock. As they grow larger, they are moved and shaped by wind.
11. An inselberg is a solid rock formation that was once below ground level. As the softer land around it erodes, it becomes more and more prominent. Uluru is an inselberg.
12. A chimney rock is the pillar-like remains of a butte.
13. Star dunes are produced by wind gusts that swirl in from all directions.
14. Strong winds blowing in one direction form longitudinal dunes.

4.5.2 Sand dunes = depositional landforms

Different dune shapes are created by the action of the wind (see **FIGURE 2**). These include crescent, linear, star, dome and parabolic. The most common are the crescent-shaped dunes that are formed when the wind blows in one direction (**FIGURE 3**). They are usually wider than they are long and can move very quickly across desert landscapes.

Linear dunes are a series of dunes running parallel to each other. They can vary in length from a few metres to over 100 kilometres. It appears that winds blowing in opposite directions help create these dunes. The Simpson Desert in central Australia has linear dunes (**FIGURE 4**).

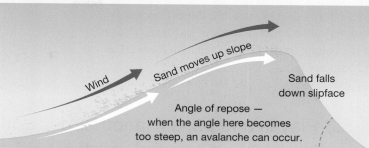

FIGURE 2 The transport and deposition of sand creates and moves dunes.

Wind

Sand moves up slope

Sand falls down slipface

Angle of repose — when the angle here becomes too steep, an avalanche can occur.

FIGURE 3 A series of crescent dunes in Egypt

FIGURE 4 Linear sand dunes in the Simpson Desert, Australia

Star dunes have 'arms' that radiate from a high central pyramid-shaped mound (**FIGURE 5**). They form in regions that have winds blowing in many different directions and can become very tall rather than wide — some are up to 500 metres high.

Dome dunes are made up of fine sand without a steep side. These rounded structures tend to be only one or two metres high and are very rare (**FIGURE 6**).

FIGURE 5 Star dunes are found in many deserts including the Namib, the Grand Erg Oriental of the Sahara, and the south-east Badain Jaran Desert of China.

FIGURE 6 A dome dune in the Chihuahuan Desert, North America

Parabolic dunes have a U shape and do not get very high (**FIGURE 7**). They often occur in coastal deserts. The longer section follows the 'head' of the dune (the opposite process to the formation of crescent dunes) because vegetation has anchored them in place. The arms can be long — in one case, measured at 12 kilometres.

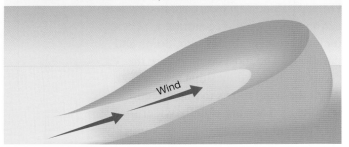

FIGURE 7 Formation of a parabolic dune

Wind

4.5.3 Playas and pans = another depositional landform

A desert basin may fill with water after heavy rains to form a shallow lake, but for the majority of the time the often salt-encrusted surface is hard and dry. Such expanses of land are known as playas, saltpans or hardpans. The flat terrains of pans and playas make them excellent race tracks and natural runways for aeroplanes and spacecraft. Ground-vehicle speed records are commonly established on Bonneville Speedway, a race track on the Great Salt Lake hardpan (**FIGURE 8**). Space shuttles land on Rogers Lake playa at Edwards Air Force Base in California in the western United States.

FIGURE 8 A driver lying in a streamlined racing car, Bonneville Salt Flats, Utah, the United States

4.5 INQUIRY ACTIVITIES

1. Locate all the desert *places* named in this subtopic. Use Google Maps to create your own map of these locations, and add some interesting facts and images of each location. Email a link to your completed map to your teacher. **Classifying, organising, constructing**

2. Draw up a table like the one below.

Name of land form	Picture of land form	Location	Type of erosion (wind or water)	Type of deposition (wind or water)
Butte				
Mesa				
Inselberg				

Continue to add the landforms shown in **FIGURE 1** to your table. Add examples of other desert landforms that you have found when researching this topic. **Classifying, organising, constructing**

3. Work in small groups to create a model of a desert (using plasticine or playdoh, for example) that contains a number of desert forms and patterns. Use **FIGURE 1** as a guide. Show your completed model to the other groups, then provide and respond to constructive feedback. **Classifying, organising, constructing**

4.5 EXERCISES

Geographical skills key: GS1 Remembering and understanding **GS2** Describing and explaining **GS3** Comparing and contrasting **GS4** Classifying, organising, constructing **GS5** Examining, analysing, interpreting **GS6** Evaluating, predicting, proposing

4.5 Exercise 1: Check your understanding

1. **GS1** List the agents of erosion and weathering in a desert. How does each process cause *change* in a desert?
2. **GS1** Name two erosional and two depositional landforms in a desert.
3. **GS1** Name the most common dune shapes that are formed in deserts.
4. **GS2** Explain the difference between a mesa and a butte.
5. **GS2** How does vegetation help to prevent erosion in a desert?
6. **GS1** What wind conditions are needed to create a:
   (a) star dune
   (b) longitudinal dune
   (c) parallel dune?

4.5 Exercise 2: Apply your understanding

1. **GS2** Why do you think oases are such fertile *places*?
2. **GS3** What do chimney rocks and arches have in common?
3. **GS3** What do playa lakes and saltpans have in common?
4. **GS6** Study the landforms labelled 1, 3 and 9 in **FIGURE 1**. Sketch what each of these may look like in the future as erosion and weathering continue to occur.
5. **GS1** Which desert in Australia contains many linear dunes?
6. **GS1** On which landforms are land-speed records often held and why?

Try these questions in learnON for instant, corrective feedback. Go to www.jacplus.com.au.

4.6 Characteristics of Australia's deserts

4.6.1 The location of Australia's deserts

Australia is the world's driest inhabited continent. Over 70 per cent of the country receives between 100 and 350 millimetres of rainfall annually, which makes most of Australia arid or semi-arid.

Australia's deserts are subtropical and are located mainly in central and western Australia, making up about 18 per cent of the country (see **FIGURE 1**). They are hot deserts, which means they are areas of little rainfall and extreme temperatures — rainfall can be less than 250 millimetres per year and temperatures can rise to over 50 °C. The average humidity is between 10 and 20 per cent. The desert terrain is very diverse and can range from red sand dunes to the polished stones of the gibber plains — the term *gibber* comes from an Aboriginal language word for stone.

Great Victoria Desert

The Great Victoria Desert, Australia's largest, covers 424 400 square kilometres. It is not a

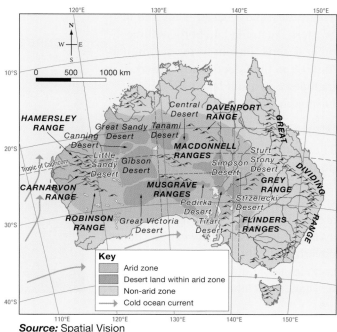

FIGURE 1 The location and distribution of Australia's deserts

Source: Spatial Vision

desert of dunes, but has some desert-adapted plants including marble gums, mulga and spinifex grass. Part of this desert has been named a Biosphere Reserve by UNESCO and is one of the largest arid zone biospheres in the world.

Great Sandy Desert

The Great Sandy Desert makes up 3.5 per cent of Australia. The red sands of this desert reach almost to the Western Australia coast, where they join with the white sand of Eighty Mile Beach south of Broome.

Simpson Desert

The Simpson Desert is in one of the driest areas of Australia, with rainfall of less than 125 millimetres per year. It is located near the geographical centre of Australia. Dunes (see **FIGURE 2**) make up nearly three-quarters of the desert. Long parallel dunes (see **FIGURE 4** in subtopic 4.5) form in a north–north-west/south–south-east direction; some can be straight and unbroken for up to 300 kilometres and can be 40 metres high. The space between the dunes can vary from 100 metres to 1000 metres.

FIGURE 2 Sand dunes and vegetation in the Simpson Desert

Strzelecki Desert

This desert is located within three states — far northern South Australia, south-west Queensland and western New South Wales. The dunes support vegetation such as sandhill wattle, needlebush and hard spinifex.

Tanami Desert

Located to the east of the Great Sandy Desert, this desert is mostly characterised by red sand plains with hills and ranges.

Little Sandy Desert

The Little Sandy Desert is located in Western Australia and borders three other deserts. Its landforms are similar to those in the Great Sandy Desert. It includes a vast salt lake called Lake Disappointment.

FIGURE 3 Gibber landscape in the Sturt Stony Desert in South Australia

Sturt Stony Desert

The Sturt Stony Desert, located in north-eastern South Australia, is a harsh gibber desert covered in closely spaced glazed stones (**FIGURE 3**). These are left behind when the wind blows away the loose sand between the dense covering of pebbles. The desert also contains some dunes and hills that are resistant to weathering.

Tirari Desert

This small desert covers almost 1600 square kilometres and is located in far northern South Australia, east of Lake Eyre. It contains many linear (parallel) dunes and salt lakes. Cooper Creek runs through the centre of the desert, as do many other **intermittent creeks**. Where there is enough water — usually in waterholes — river red gums and coolabah gums will grow. Tall, open shrubland also occurs in some areas.

Gibson Desert

The fifth largest in Australia, the Gibson Desert is located in Western Australia and borders three other deserts. It consists of sand plains and dunes plus some low, rocky ridges. Some small salt-water lakes are also present in the south-western part of the desert.

FIGURE 4 Desert between Oodnadatta and William Creek, South Australia

Pedirka Desert

The Pedirka Desert in South Australia is Australia's smallest desert, located north-east of Oodnadatta. The lines of parallel red dunes run north-east to south-west, and the space between the dunes can be up to one kilometre. Hamilton Creek is located in this desert and its banks are home to river red gums, coolabah, mulga and prickly wattle. Other vegetation includes satiny bluebush, weeping emubush and spiny saltbush. Common grasses include woollybutt, broad-leaf wanderrie, mulga grass and bandicoot grass.

 Resources

 **Weblink** Meteorology

---

4.6 INQUIRY ACTIVITIES

1. Research the characteristics of the Biosphere Reserve declared by UNESCO that is located in the Great Victorian Desert. **Examining, analysing, interpreting**
2. Use an atlas to find the locations of Brisbane, Geraldton and Exmouth. These *places* are located at the same latitude as many of Australia's deserts. Use the **Meteorology** weblink in the Resources tab to find the average temperature, rainfall and humidity of these *places*.
   (a) How do these characteristics compare with the temperature, rainfall and humidity in Australia's deserts?
   (b) How can you account for the differences? **Comparing and contrasting**
3. Several plants are listed in the descriptions in this subtopic on Australia's deserts. Choose two different plant types (for example, a grass and a tree) and research how they are adapted to desert conditions.
   Examining, analysing, interpreting

4.6 EXERCISES

Geographical skills key: GS1 Remembering and understanding **GS2** Describing and explaining **GS3** Comparing and contrasting **GS4** Classifying, organising, constructing **GS5** Examining, analysing, interpreting **GS6** Evaluating, predicting, proposing

4.6 Exercise 1: Check your understanding

1. **GS1** Name the deserts bordered by:
   (a) the Gibson Desert
   (b) the Little Sandy Desert.

▶

---

2. **GS1** What is a gibber desert?
3. **GS1** What percentage of Australia is arid or semi-arid?
4. **GS1** Where is the Great Victoria desert located?
5. **GS1** Name Australia's smallest desert. Where is it located?

4.6 Exercise 2: Apply your understanding

1. **GS2** Look at **FIGURE 1** showing the distribution of Australia's deserts. Where are they located in terms of the tropics?
2. **GS1** List the types of vegetation that can be found in the Strzelecki Desert.
3. **GS1** Which desert is Australia's driest and what are its characteristics?
4. **GS1** What is an intermittent creek?
5. **GS1** Which desert contains a UNESCO Biosphere Reserve?

Try these questions in learnON for instant, corrective feedback. Go to www.jacplus.com.au.

4.7 SkillBuilder: Calculating distance using scale

What does it mean to calculate distance using scale?

Calculating distance using scale involves working out the actual distance from one place to another using a map. The scale on a map allows you to convert distance on a map or photograph to distance in the real world. A linear scale is the easiest to use.

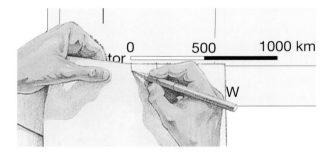

Select your learnON format to access:

- an overview of the skill and its application in Geography (Tell me)
- a video and a step-by-step process to explain the skill (Show me)
- an activity and interactivity for you to practise the skill (Let me do it)
- questions to consolidate your understanding of the skill.

 Resources

🎞 **Video eLesson** SkillBuilder: Using latitude and longitude (eles-1652)

🧩 **Interactivity** SkillBuilder: Using latitude and longitude (int-3148)

4.8 How did Lake Mungo become dry?

4.8.1 Where are Lake Mungo and the Willandra Lakes located?

Lake Mungo, in Mungo National Park, is just one of 13 ancient dry lake beds in a section of the Willandra Lakes Region World Heritage area in semi-arid New South Wales. There is no water there now, yet the lakes were once full of water and teeming with life, supporting Aboriginal peoples since the beginning of the Dreamings (more than 47 000 years by European estimates) — archaeological records show this continuous human presence. What happened to change this environment into the semi-arid landscape it is today?

The Willandra Lakes are located in far south-western New South Wales and the region is part of the Murray–Darling River Basin. Lake Mungo is 110 kilometres north-east of Mildura, Victoria. The lakes were originally fed by water from Willandra Creek (see **FIGURE 1**), which was a branch of the Lachlan River. The average rainfall in this area is 325 millimetres per year, making it a semi-arid desert region.

4.8.2 How has Lake Mungo changed over time?

40 000 years ago

During the last ice age, huge amounts of water filled the shallow lake. At its fullest, Lake Mungo was 6–8 metres deep and covered 130 square kilometres (more than twice the area of Sydney Harbour). The lakes were rich with life, including water birds, freshwater mussels, yabbies and fish such as golden perch and Murray cod. Giant kangaroos, giant wombats, large emus and the buffalo-sized Zygomaturus — all now extinct — grazed around the water's edge. Remains of more than 55 species have been found in the area and identified — 40 of these are no longer found in the region, and 11 are extinct.

Aboriginal peoples lived here in large numbers — evidence for this has been found in more than 150 human fossils, including 'Mungo Lady' discovered in 1968 and 'Mungo Man' in 1974, both believed to be over 40 000 years old. The youngest fossil is 150 years old.

FIGURE 1 Location of Willandra Lakes, including Lake Mungo

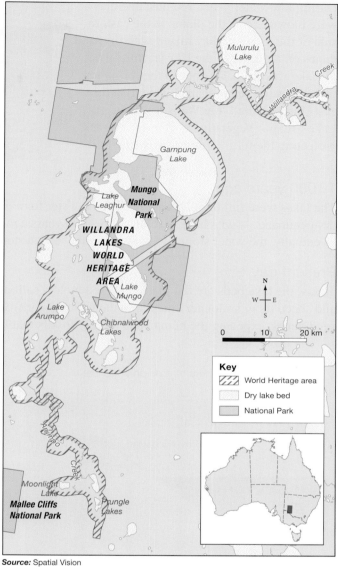

Source: Spatial Vision

FIGURE 2 Traditional owners clean fossilised footprints at Lake Mungo.

DISCUSS

'It is right for Lake Mungo to be protected under a World Heritage listing because of its significant cultural characteristics.' How does this type of protection reflect the cultural values of a society?

[Intercultural Capability]

30 000–19 000 years ago

A west wind blows across this landscape. During low-water years, red dust and clay were blown across the plains to the eastern side of the lake and they mixed with the sand dunes on the edge of the lake (formed when the lake was full). This began the formation of lunettes (crescent-shaped dunes) on the east side — called 'the Walls of China' in Lake Mungo. Vegetation covered the dunes, protecting them.

FIGURE 3 The 'Walls of China' at Lake Mungo. The dry lake bed is covered by low bushes and grasses.

19 000 years ago

The lakes were full of deep, relatively fresh water for a period of 30 000 years — with cycles of wet and dry occurring — which came to an end 19 000 years ago when the climate became drier and warmer. Eventually, the water stopped flowing into the lake system and it dried out.

Present day

Today, the lake beds are flat plains covered by low saltbush and bluebush as well as grasses. Grazing cattle and sheep (now no longer allowed in the national park) and rabbits have caused erosion of the lunettes and sand dunes, exposing the human and animal fossils that have since been discovered.

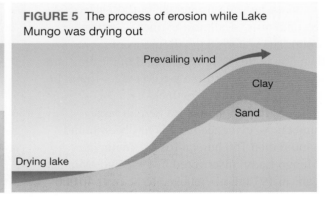

FIGURE 4 The process of erosion while Lake Mungo was full

FIGURE 5 The process of erosion while Lake Mungo was drying out

4.8.3 World Heritage listing

The Willandra Lakes Region, which includes Lake Mungo, is listed as a World Heritage Area. This region is important because of its archaeology (human skeletons, tools, shell middens and animal bones make up the oldest evidence of burial places in the world) and geomorphology (ancient and undisturbed landforms and sediments).

on Resources

Interactivity Evolving Lake Mungo (int-3108)

Weblink Timeline maker

4.9 How people use deserts

4.9.1 Traditional livelihoods

Although not many people live in deserts, these environments have been important to traditional communities for many years. People either adapt to living in deserts or transform deserts to suit their needs. People are also attracted to desert regions to mine resources.

There are many communities who live in deserts including Indigenous Australian peoples; the Bedouin people of the Middle East and Sahara; the Tuareg people of the Sahara in North Africa; the Topnaar people in the Namib Desert; the San people in South Africa; the Timbisha Shoshone of the Mojave Desert in the United States; and the communities from the Atacama Desert in South America.

Many of these communities are **nomadic**, moving with the seasons and obtaining all their needs from the land or herding animals and trading with people in settlements. It is important to understand that not all desert peoples are desert-dwellers. For example, many Indigenous Australian people do not live in deserts.

FIGURE 1 The San people in southern Africa

4.9.2 Desert resources

Many of the changes in deserts have been brought about by developments in technology. These changes have resulted in water being extracted and used to grow crops, and minerals being mined and used in many ways.

Water in the desert

Drilling equipment and pumps have allowed deep bores to tap into groundwater in aquifers deep below the desert surface. This has transformed some deserts in northern Africa and the Middle East into a series of circular irrigation fields — some of these can be up to three kilometres in diameter. In Australia, groundwater from the Great Artesian Basin has enabled desert communities to exist and grazing to take place. Unfortunately, the groundwater in many areas is being pumped out far more quickly than it is being replaced and may be in danger of running out.

Desalination plants have also provided water to desert communities in many areas, especially the Middle East, including large cities such as Dubai.

FIGURE 2 (a) Satellite image of circular irrigation fields in Libya; (b) aerial photograph of a circular irrigation field

Mining in deserts

Many deserts contain valuable mineral deposits that were formed in the arid environment or have been exposed by erosion. Desert mining has created a lot of wealth for some people and companies, but usually not for the traditional desert people. Examples of mining resources include:

1. iron and lead-zinc ore — mined in Australian deserts
2. phosphorus (used to make fertilisers) — mined in the Sahara region
3. borates (used to manufacture glass, ceramics, enamels and agricultural chemicals) — mined in the deserts of California, United States
4. copper, iron ore and nitrates — mined in Chile's Atacama Desert
5. precious metals such as gold, silver and platinum — mined in the deserts of Australia, the United States and central Asia
6. uranium — mined in Australia and the United States
7. diamonds — mined in the Kalahari and Namib deserts of south-western Africa
8. oil — more than 65 per cent of the world's oil is found in the desert regions of the Middle East, mainly in Kuwait, Iraq, Iran and Saudi Arabia.

FIGURE 3 A uranium mine next to the Colorado River in the United States

4.9 EXERCISES

Geographical skills key: GS1 Remembering and understanding **GS2** Describing and explaining **GS3** Comparing and contrasting **GS4** Classifying, organising, constructing **GS5** Examining, analysing, interpreting **GS6** Evaluating, predicting, proposing

4.9 Exercise 1: Check your understanding

1. **GS1** List the sources of water in a desert that can be used to grow crops and provide water for people.
2. **GS2** How has technology enabled water to be used in deserts?
3. **GS2** Do you think most desert people adapt to live in the desert *environment*, or adapt the *environment* to live in the desert? Give two examples to support your reasoning.
4. **GS2** Why is it important to use groundwater *sustainably*?
5. **GS1** Where do the following indigenous peoples live: Bedouin, Tuareg, Topnaar, San and Timbisha?

4.9 Exercise 2: Apply your understanding

1. **GS5** Study **FIGURE 2a**.
   (a) Where is Libya located? Use an atlas or Google Earth and write a description.
   (b) Identify the small red circles in the image.
2. **GS6** Study **FIGURE 3**.
   (a) Make a sketch of **FIGURE 3**. Label your sketch to include the river, the mine site and the buildings.
   (b) How has mining *changed* this *environment*?
   (c) What issues could arise due to the location of this mine?
   (d) Predict what might happen to this area when mining stops.
3. **GS6** What is meant by the term 'nomadic'? Why might a desert *environment* suit the needs of nomadic peoples?
4. **GS1** Where is the Great Artesian Basin located and why is it important?
5. **GS6** Suggest specific ways that desert mining might affect desert-dwelling people.

Try these questions in learnON for instant, corrective feedback. Go to www.jacplus.com.au.

4.10 Antarctica — a cold desert

4.10.1 Some facts about Antarctica

Like hot deserts, polar deserts are areas with annual precipitation of less than 250 millimetres, but they have a mean temperature during the warmest month of less than 10 °C. Polar deserts are found in both the Arctic and Antarctic regions of the world. Not only is Antarctica a desert — it is also the driest, coldest and windiest continent on Earth.

Australia is the driest inhabited continent on Earth. However, Antarctica is even drier. Much of Australia's interior receives less than 250 millimetres of precipitation per year. The interior of Antarctica receives less than 50 millimetres. The coastal areas receive the highest levels of precipitation, but this is still only about 200 millimetres.

Most of Antarctica is too cold for rainfall; the majority of the precipitation falls as snow. Some valleys in Antarctica have received no rain for two million years. It also snows very little in Antarctica, particularly in the interior.

In places, the ice sheet in Antarctica is 4.8 kilometres deep (see **FIGURE 1**). Most of the ice that covers the continent has been there for thousands of years. In winter, as the surrounding oceans freeze, the area of Antarctica is almost double that in summer.

FIGURE 1 This cross-section, which shows the mountains below the ice, passes through some of the thickest parts of the Antarctic ice sheet.

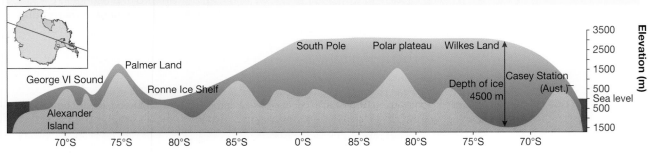

How dry is dry?

Covered in ice, Antarctica may seem like the wettest place in the world, but it's actually drier than the Sahara Desert. Despite this, Antarctica's ice holds 70 per cent of the world's fresh water supply.

Most places in Antarctica receive no rain or snow at all. Very cold air does not have the capacity to hold enough water to create rain or snow. This means that Antarctica is the world's biggest desert. All drinking water in Antarctica is obtained by melting the ice. Unlike in hot deserts, there is little evaporation from Antarctica, so the relatively small amount of snow that does fall doesn't disappear. Instead it builds up over hundreds and thousands of years into enormously thick ice sheets.

FIGURE 2 A climograph for Mawson Base, Antarctica

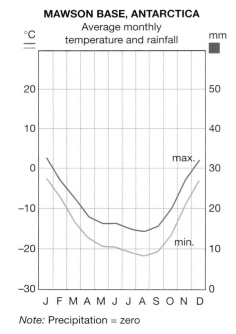

Note: Precipitation = zero

How cold is cold?

During the winter of 2018, NASA scientists used satellites to record a temperature of –98.7 °C, the coldest ever recorded on Earth. The measurements were taken in a depression on one of the highest points of the dome-shaped ice sheet. During the coldest months (July to August), the average temperature at the South Pole is –60 °C. During the warmest months (December to January), it rises to –28 °C.

Why is Antarctica so cold?

There are three main reasons:

1. Antarctica's position on the globe means that the sun's rays strike the Earth's surface at a low angle, and therefore have a much larger area to heat than at other places on the planet.
2. Most of the sun's heat that does reach Antarctica is reflected back into space by the white ice that covers the continent. This also explains why you must always wear sunglasses or goggles in Antarctica.
3. Antarctica is surrounded by the cold waters of the Southern Ocean.

How windy is windy?

Australia's greatest polar explorer, Douglas Mawson, called Antarctica 'the home of the blizzard'. He should know. He lived in a wooden hut for two complete Antarctic winters, in the strongest winds ever recorded. Mawson's measurements revealed an average wind speed of over 70 kilometres per hour and gusts of over 300 kilometres per hour! The men in his expedition team always carried an ice axe with them to avoid being blown into the sea.

Why is Antarctica so windy?

As the air over the polar plateau becomes colder, it becomes more dense. Finally gravity pulls it down off the plateau towards the Antarctic coast. This creates very strong winds, called **katabatic winds**, which can blow continually for weeks on end and carry small pellets of ice. These winds combined with the severe cold can be fatal; at −20 °C, exposed human flesh begins to freeze when the wind reaches only 14 kilometres per hour.

Katabatic winds also cause **blizzards**, which sweep up loose snow and blow it about ferociously. Such blizzards were the cause of death among many early Antarctic explorers.

The winds also shape the landscape, carving it into irregular shapes called **sastrugi** (see **FIGURE 3**). These shapes range in height from 150 millimetres to two metres. Travelling across sastrugi is extremely difficult.

FIGURE 3 Sastrugi lie in the direction of the prevailing wind. They are as hard as rock.

4.10.2 How do people use Antarctica?

Antarctica and the seas that surround it contain valuable resources. Antarctica is also the temporary home of more than 4000 people in summer and 1000 in winter. Most are scientists and support staff. These people work in more than 66 research stations representing 30 different nations.

Mining

There are great difficulties in looking for mineral deposits in rocks that lie beneath thousands of metres of ice. Therefore, most exploration has taken place in the ice-free areas of Antarctica. Scientists now believe there are deposits of many valuable minerals in Antarctica, including coal, iron ore, copper, lead and uranium, and traces of minerals such as gold and zinc. There are also mineral beds lying under the continent's Transantarctic Mountains, and large areas that may contain deposits of oil and gas (see **FIGURE 5**).

FIGURE 4 Esperanza, a permanent, all-year round Argentinian research base, Graham Land, Antarctica

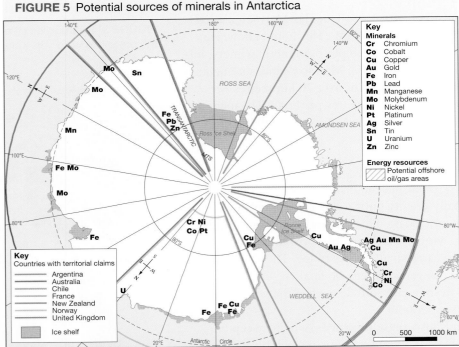

FIGURE 5 Potential sources of minerals in Antarctica

Source: © Geography Teachers' Association of Victoria, Inc.

Despite the presence of these valuable minerals, there are no operating mines in Antarctica. Given the conditions — the extreme cold, the rough seas and the wind — mining operations would be very difficult and potentially dangerous to the environment. Mining (other than for scientific purposes) is banned under the Antarctic Treaty. This is to prevent the possibility of polluting the environment (for example, through an oil spill or by digging a quarry).

The Antarctic Treaty

By the mid-1950s, Australia, New Zealand, the United Kingdom, France and Argentina were actively exploring Antarctica. These countries declared territorial claims over parts of Antarctica while others were fishing, whaling and conducting scientific research and mineral exploration in the region.

People began to realise that this unique wilderness needed to be protected. In 1958, 12 countries agreed to preserve Antarctica. This led to an international agreement called the Antarctic Treaty, which came into force in 1961. The **treaty** covers the area south of 60 °S latitude. It has been signed by more than 52 countries who meet regularly to discuss issues affecting Antarctica. The treaty:

- prohibits military activity
- protects the Antarctic environment
- fosters scientific research
- recognises the need to protect Antarctica from uncontrolled destruction and interference by people.

4.10.3 Tourism

The number of tourists to Antarctica has increased significantly since the mid-1990s, with a peak of over 51 700 in 2017–18. However, more people will attend one game of AFL football in Melbourne than will visit Antarctica in one year. Given the scale (size) of Antarctica, tourist numbers are therefore still small but continue to increase each year.

Most tourists go to Antarctica on board cruise ships. There are opportunities for people to land on the ice. This often requires use of a Zodiac inflatable boat between ship and shore. There are no tourist facilities on Antarctica — people must return to the ships, for example, to sleep, eat and shower.

Sightseeing is the main activity for tourists. Other activities include kayaking, visiting research stations, walking and snowboarding. Other types of tourism include flights over the continent and flights that include landing on the ice.

Tourism can create problems, such as pollution from oil spills and disturbance to animal colonies. Therefore, the International Association of Antarctica Tour Operators has set up rules to control tourism. For example, no more than 100 passengers from a cruise ship may be landed at a location in Antarctica at any one time.

TABLE 1 Tourists to Antarctica

Year	Tourist numbers	Year	Tourist numbers
1996–97	7330	2007–08	46 069
1997–98	9604	2008–09	37 858
1998–99	10 013	2009–10	36 975
1999–2000	14 762	2010–11	33 824
2000–01	12 248	2011–12	26 509
2001–02	11 588	2012–13	34 354
2002–03	13 571	2013–14	37 405
2003–04	27 537	2014–15	36 702
2004–05	27 950	2015–16	38 478
2005–06	29 823	2016–17	44 367
2006–07	29 823	2017–18	51 707

Source: International Association of Antarctica Tour Operators

Bases on ice

Most of Antarctica's scientific bases are located on the coast so people and supplies can be brought in by boat or air (see **FIGURE 6**). They are also situated on the two per cent of Antarctica not covered in ice, as bases built on ice tend to sink under their own weight. This is because the heat they generate can melt ice around and beneath them.

Some bases are inland. There is even a permanent scientific base at the South Pole: the American Amundsen–Scott Base. Australia operates three permanent bases in Antarctica — Casey, Mawson and Davis stations — plus one on Macquarie Island and five temporary summer bases.

In January 2008 an air link between Australia and Antarctica was officially opened. The Wilkins runway is a four-kilometre-long airstrip about 70 kilometres from Casey Station. Scientists can now get to Antarctica in a few hours from Australia rather

FIGURE 6 Davis is the most southerly Australian Antarctic station.

Source: Australian Antarctic Division

than a few weeks on a ship. They can study the world's weather, climate, marine and land biology, glaciers, magnetics, geology and the ozone layer, as well as human physiology. Ice cores can provide a record of climate change over a long period of time.

Explore more with my**World**Atlas

Deepen your understanding of this topic with related case studies and questions.
- Exploring places > Antarctica > **Antarctica: human features**

 Resources

> **Weblinks**   Antarctic weather
>   Meteorology
>   Life in Antarctica
>   Antarctic
>   Biosecurity fears

4.10 INQUIRY ACTIVITIES

1. Use the **Antarctic weather** and **Meteorology** weblinks in the Resources tab to describe the weather conditions now at the South Pole. Compare these to the conditions where you live.
   Comparing and contrasting
2. Use an atlas to measure the distance from Antarctica (coastline) to South America, Australia and South Africa. **Examining, analysing, interpreting**
3. Use the information in **TABLE 2** below to draw a climograph of McMurdo Station. How does it compare to Mawson Base (see **FIGURE 2**)? Find climate data for the *place* where you live and draw another climograph for that location. Compare this to the two Antarctic climographs. Outline the similarities and differences and provide reasons for these. **Comparing and contrasting**

TABLE 2 Climate data for the American McMurdo station in Antarctica: latitude 77.88°S, longitude 166.73°E

	Jan	Feb	Mar	Apr	May	Jun	Jul	Aug	Sep	Oct	Nov	Dec	Annual
Average daily temperature (°C)	−2.9	−9.5	−18.2	−20.7	−21.7	−23.0	−25.7	−26.1	−24.6	−18.9	−9.7	−3.4	Mean −16.9
Mean monthly rainfall (mm)	15	21.2	24.1	18.4	23.7	24.9	15.6	11.3	11.8	9.7	9.5	15.7	Total 202.5

4. Working in groups of four, use the **Life in Antarctica** weblink in the Resources tab to investigate life at the Australian Antarctic stations. Choose one station.
   (a) What facilities are there at the station?
   (b) Describe the work activities that take place.
   (c) What do you think it is like to live there? **Describing and explaining**
5. Use a spreadsheet program to draw a line graph using the tourism data in **TABLE 1**. Describe how the numbers have *changed* over time and provide possible explanations for these *changes*.
   Describing and explaining
6. Use the **Antarctic** and **Biosecurity fears** weblinks in the Resources tab to find out more about how foreign seeds are invading Antarctica. Write a list of rules for a company that would remove this risk.
   Evaluating, predicting, proposing

7. Do you think countries should be able to own pieces of Antarctica? Write a two-minute speech outlining the reasons for your point of view. Debate this topic as a class. **Evaluating, predicting, proposing**

8. Would you like to visit Antarctica? Why? Discuss as a class, listening carefully to the opinions of others.
Evaluating, predicting, proposing

4.10 EXERCISES

Geographical skills key: GS1 Remembering and understanding **GS2** Describing and explaining **GS3** Comparing and contrasting **GS4** Classifying, organising, constructing **GS5** Examining, analysing, interpreting **GS6** Evaluating, predicting, proposing

4.10 Exercise 1: Check your understanding

1. **GS1** List three facts about Antarctica that you found the most surprising.
2. **GS1** Why does Antarctica double in area every winter?
3. **GS1** What is the coldest temperature ever recorded in Antarctica, and in which year was the temperature recorded?
4. **GS2** Antarctica is sometimes described as the world's biggest desert. Why?
5. **GS2** Describe and explain why Antarctica is so dry, cold and windy.
6. **GS2** Examine the photograph in **FIGURE 3** and describe how this landscape has been formed. How does this *environment* pose a risk to people?

4.10 Exercise 2: Apply your understanding

1. **GS6** What might happen to Antarctica if the ice shelves on top of the mountains were to melt? What *changes* might happen to sea levels around the world? Construct a concept map to record all your ideas.
2. **GS2** List three ways in which the stations might have an impact on the Antarctic *environment*.
3. **GS2** Why is there no mining in Antarctica? What problems would there be in extracting and transporting minerals from Antarctica?
4. **GS6** Suggest the ideal location for a scientific base in Antarctica.
5. **GS2** Why don't tourists visit Antarctica during winter?

Try these questions in learnON for instant, corrective feedback. Go to www.jacplus.com.au.

4.11 Thinking Big research project: Desert travel brochure

Online only

SCENARIO

Your graphic design business is applying for a rewarding contract to design travel brochures to amazing locations. The brief you have been given for the job interview is to create a brochure enticing people to visit one of the world's deserts.

Select your learnON format to access:

- the full project scenario
- details of the project task
- resources to guide your project work
- an assessment rubric.

 Resources

projectsPLUS Thinking Big research project: Desert travel brochure (pro-0170)

4.12 Review

4.12.1 Key knowledge summary

Use this dot point summary to review the content covered in this topic.

4.12.2 Reflection

Reflect on your learning using the activities and resources provided.

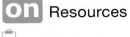 Resources

> **eWorkbook** Reflection (doc-31348)
>
> Crossword (doc-31349)
>
> **Interactivity** Desert landscapes crossword (int-7597)

KEY TERMS

blizzard a strong and very cold wind containing particles of ice and snow that have been whipped up from the ground

humidity the amount of water vapour in the atmosphere

intermittent creek a creek that flows for only part of the year following rainfall

katabatic wind very strong winds that blow downhill

nomadic describes a group that moves from place to place depending on the food supply, or pastures for animals

rain shadow the drier side of a mountain range, cut off from rain-bearing winds

sastrugi parallel wave-like ridges caused by winds on the surface of hard snow, especially in polar regions

treaty a formal agreement between two or more countries

5 Mountain landscapes

5.1 Overview

Magma, water and tectonic plates — can they really move mountains?

5.1.1 Introduction

Mountains occupy 24 per cent of the Earth's landscape, and are characterised by many different landforms. The forces that form and shape mountains come from deep within the Earth, and have been shaping landscapes for millions of years. The Earth is a very active planet — every day, many volcanoes are erupting somewhere on the planet, and even more tremors are occurring. In this topic we will explore the mountains of the world, how they are formed and the ways that people use them. We will also look at earthquakes, tsunamis and volcanoes, and the effects they have on people and places.

On Resources

☑ **eWorkbook**   Customisable worksheets for this topic

▦ **Video eLesson** Majestic mountains (eles-1626)

LEARNING SEQUENCE

5.1 Overview
5.2 How mountains are formed
5.3 The world's mountains and ranges
5.4 **SkillBuilder:** Drawing simple cross-sections   `online only`
5.5 How people use mountains
5.6 Earthquakes and tsunamis
5.7 Volcanic mountains
5.8 **SkillBuilder:** Interpreting an aerial photo   `online only`
5.9 How do volcanic eruptions affect people?
5.10 **Thinking Big research project:** Earthquakes feature article   `online only`
5.11 **Review**   `online only`

To access a pre-test and starter questions and receive immediate, **corrective feedback** and **sample responses** to every question, select your learnON format at www.jacplus.com.au.

5.2 How mountains are formed

5.2.1 What are the forces that form mountains?

A mountain is a landform that rises high above the surrounding land. Most mountains have certain characteristics in common, although not all mountains have all these features. Many have steep sides and form a peak at the top, called a summit. Some mountains located close together have steep valleys between them known as gorges.

Mountains and mountain ranges have formed over billions of years from tectonic activity; that is, movement in the Earth's crust. The Earth's surface is always changing — sometimes very slowly and sometimes dramatically.

Continental plates

The Earth's crust is cracked and is made up of many individual moving pieces called continental plates, which fit together like a jigsaw puzzle. These plates float on the semi-molten rocks, or magma, of the Earth's mantle. Enormous heat from the Earth's core, combined with the cooler surface temperature, creates **convection currents** in the magma. These currents can move the plates by up to 15 centimetres per year. Plates beneath the oceans move more quickly than plates beneath the continents.

Continental drift

Scientific evidence shows that about 225 million years ago all the continents were joined.

FIGURE 1 World map of plates, volcanoes and hotspots

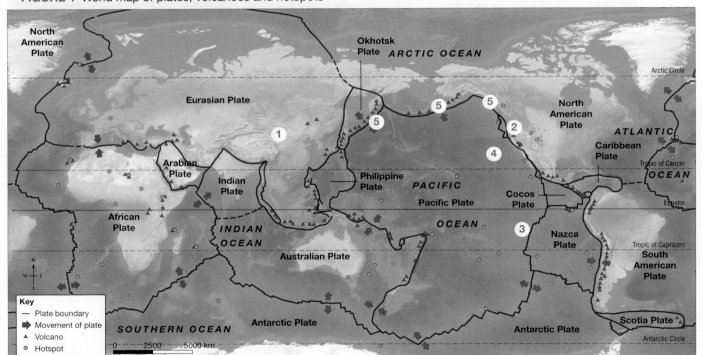

Source: Map drawn by Spatial Vision

① Convergent plates

When two continental plates of similar density collide, the pressure of the **converging plates** can push up land to form mountains. The Himalayas were formed by the collision of the Indian subcontinent and Asia. The European Alps were formed by the collision of Africa and Europe.

When an oceanic and a continental plate collide, they are different densities, and the thinner oceanic plate is subducted, meaning it is forced down into the mantle. Heat melts the plate and pressure forces the molten material back to the surface. This can produce volcanoes and mountain ranges. The Andes in South America were formed this way.

Subduction can also occur when two oceanic plates collide. This forms a line of volcanic islands in the ocean about 70–100 kilometres past the subduction line. The islands of Japan have been formed in this way. Deep oceanic trenches are also formed when this occurs. The Mariana Trench in the Pacific Ocean is 2519 kilometres long and 71 kilometres wide, and is the deepest point on Earth — 10.911 kilometres deep. If you could put Mount Everest on the ocean floor in the Mariana Trench, its summit would lie 1.6 kilometres below the ocean surface.

② Lateral plate slippage

Convection currents can sometimes cause plates to slide, or slip, past one another, forming **fault** lines. The San Andreas Fault, in California in the western United States, is an example of this.

③ Divergent plates

In some areas, plates are moving apart, or diverging, from each other (for example, the Pacific Plate and Nazca Plate). As the **divergent plates** separate, magma can rise up into the opening, forming new land. Underwater volcanoes and islands are formed in this way.

④ Hotspots

There are places where volcanic eruptions occur away from plate boundaries. This occurs when there is a weakness in the oceanic plate, allowing magma to be forced to the surface, forming a volcano. As the plate drifts over the **hotspot**, a line of volcanoes is formed.

⑤ The Pacific Ring of Fire

The most active region in the world is the Pacific Ring of Fire. It is located on the edges of the Pacific Ocean and is shaped like a horseshoe. The Ring of Fire is a result of the movement of tectonic plates. For example, the Nazca and Cocos plates are being subducted beneath the South American Plate, while the Pacific and Juan de Fuca plates are being subducted beneath the North American Plate. The Pacific Plate is being subducted under the North American Plate on its east and north sides, and under the Philippine and Australian plates on its west side. The Ring of Fire is an almost continuous line of volcanoes and earthquakes. Most of the world's earthquakes occur here, and 75 per cent of the world's volcanoes are located along the edge of the Pacific Plate.

FIGURE 2 The Earth's core is very hot, while its surface is quite cool. This causes hot material within the Earth to rise until it reaches the surface, where it moves sideways, cools, and then sinks.

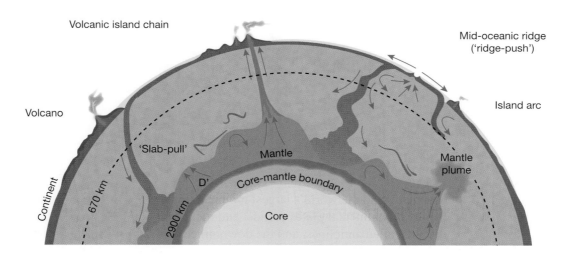

5.2.2 How do different types of mountains form?

The different movements and interactions of the **lithosphere** plates result in many different mountain landforms. Mountains can be classified into five different types, based on what they look like and how they were formed. These are fold, fault-block, dome, plateau and volcanic mountains. (Volcanic mountains are discussed in subtopic 5.7.)

FIGURE 3 Selected world mountains

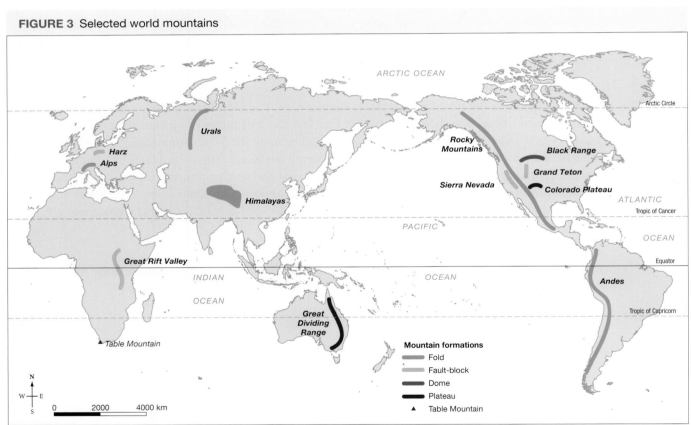

Source: Spatial Vision

Fold mountains

The most common type of mountain, and the world's largest mountain ranges, are fold mountains. The process of folding occurs when two continental plates collide, and rocks in the Earth's crust buckle, fold and lift up. The upturned folds are called anticlines, and the downturned folds are synclines (see **FIGURE 4**). These mountains usually have pointed peaks.

FIGURE 4 The formation of fold mountains

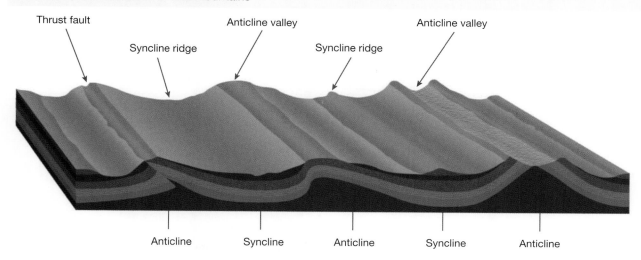

Thrust fault

Syncline ridge

Anticline valley

Syncline ridge

Anticline valley

Anticline　　Syncline　　Anticline　　Syncline　　Anticline

Examples of fold mountains include:
- the Himalayas in Asia
- the Alps in Europe
- the Andes in South America
- the Rocky Mountains in North America
- the Urals in Russia.

 **Resources**

🔗 **Weblink** Anticline and syncline

Fault-block mountains

Fault-block mountains form when faults (or cracks) in the Earth's crust force some parts of rock up and other parts to collapse down. Instead of folding, the crust fractures (pulls apart) and breaks into blocks. The exposed parts then begin to erode and shape mountains and valleys (see **FIGURE 6**).

Fault-block mountains usually have a steep front side and then a sloping back. The Sierra Nevada and Grand Tetons in North America, the Great Rift Valley in Africa, and the Harz Mountains in Germany are examples of fault-block mountains. Another name for the uplifted (upthrown) blocks is *horst*, and the collapsed (downthrown) blocks are *graben*.

FIGURE 5 A cliff overlooking the Great Rift Valley in northern Tanzania, Africa. These are examples of fault-block mountains.

FIGURE 6 The formation of fault-block mountains

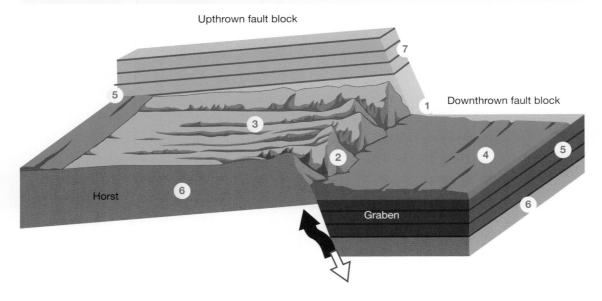

Upthrown fault block

Downthrown fault block

7

5

1

3

2

4

5

Horst

6

Graben

6

1	Fault zone	5	Sedimentary rock layers
2	Steep eastern face	6	Bedrock
3	Gentle western slope	7	Sedimentary rock layers (5) now worn away.
4	Valley floors filled with sediments of cobbles, gravel and sand		

Dome mountains

Dome mountains are named after their shape, and are formed when molten magma in the Earth's crust pushes its way towards the surface. The magma cools before it can erupt, and it then becomes very hard. The rock layers over the hardened magma are warped upwards to form the dome. Over time, these erode, leaving behind the hard granite rock underneath (see **FIGURE 7**).

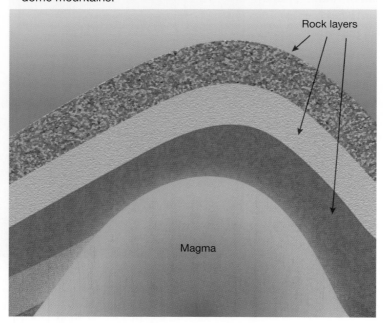

FIGURE 7 Very hot magma pushes towards the surface to form dome mountains.

Rock layers

Magma

FIGURE 8 Ben Nevis in Scotland is an example of a dome mountain.

 Resources

Google Earth Great Rift Valley
Ben Nevis

Plateau mountains

Plateaus are high areas of land that are large and flat. They have been pushed above sea level by tectonic forces or have been formed by layers of lava. Over billions of years, streams and rivers cause erosion, leaving mountains standing between valleys. Plateau mountains are sometimes known as erosion mountains.

Examples of plateau mountains include Table Mountain in South Africa (see **FIGURE 9**), the Colorado Plateau (see **FIGURE 10**) in the United States and parts of the Great Dividing Range in Australia.

FIGURE 9 The plateau of Table Mountain towers over the city of Cape Town in South Africa.

FIGURE 10 The Colorado Plateau in the United States was raised as a single block by tectonic forces. As it was uplifted, streams and rivers cut deep channels into the rock, forming the features of the Grand Canyon.

5.2.3 CASE STUDY: How were the Himalayas formed?

Before the theory of tectonic plate movement, scientists were puzzled by findings of fossilised remains of ancient sea creatures near the Himalayan peaks. Surely these huge mountains could not once have been under water?

Since understanding plate movements, the mystery has been solved. About 220 million years ago, India was part of the ancient supercontinent we call **Pangaea**. When Pangaea broke apart, India began to move northwards at a rate of about 15 centimetres per year. About 200 million years ago, India was an island separated from the Asian continent by a huge ocean.

When the plate carrying India collided with Asia 40 to 50 million years ago, the oceanic crust (carrying fossilised sea creatures) slowly crumpled and was uplifted, forming the high mountains we know today. It also caused the uplift of the Tibetan Plateau to its current position. The Bay of Bengal was also formed at this time.

The Himalayas were therefore formed when India crashed into Asia and pushed up the tallest mountain range on the continents.

The Himalayas are known as young mountains, because they are still forming. The Indian and Australian plates are still moving northwards at about 45 millimetres each year, making this boundary very active. It is predicted that over the next 10 million years it will travel more than 180 kilometres into Tibet and that the Himalayan mountains will increase in height by about five millimetres each year. Old mountains are those that have stopped growing and are being worn down by the process of erosion.

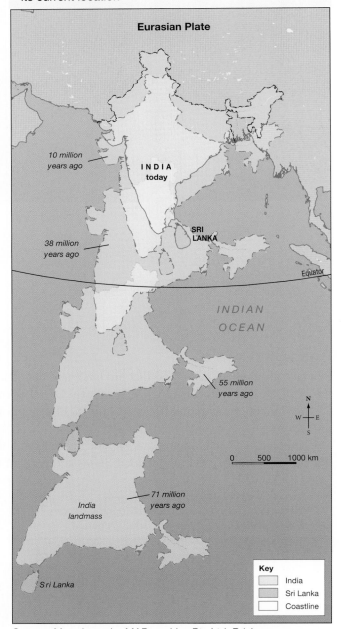

FIGURE 11 The movement of the Indian landmass to its current location

Source: Map drawn by MAPgraphics Pty Ltd, Brisbane

5.2.4 CASE STUDY: Formation of the Sierra Nevada Range, United States

The Sierra Nevada Range began to rise about five million years ago. As the western part of the block tilted up, the eastern part dropped down. As a result there is a long, gentle slope towards the west and a steep slope to the east.

FIGURE 12 The Sierra Nevada Range was formed by fault-block tilting.

East

Mono Basin

SIERRA NEVADA

Central Valley of California

West

Fault

FIGURE 13 Yosemite Valley in the Sierra Nevada mountains

on Resources

Interactivity Grand peaks (int-3110)

Weblink Fold mountains

Google Earth Sierra Nevada mountains

Table Mountain

Grand Canyon

5.2 INQUIRY ACTIVITIES

1. Use different coloured strips of plasticine to make models showing how a collision of continental and oceanic plates differs from a collision of two continental plates. Have a go at explaining this to a member of your family.   **Classifying, organising, constructing**

2. Use the **Fold mountains** weblink in the Resources tab to explain the formation of fold mountains and fault-block mountains.   **Describing and explaining**

3. Sketch **FIGURE 5** and annotate it to show where erosion has taken place. Label *places* that have hard and weak rocks.   **Describing and explaining**

4. Draw a sketch of **FIGURE 13**, noting the plateau and areas of erosion and weathering.
                                                                                                Describing and explaining

▶

5. Use an atlas to locate the Sierra Nevada Range. Describe where it is. Name two national parks in this mountain range. Choose one, and investigate some of its geographical characteristics. Present this as a PowerPoint, Keynote or Prezi presentation. **Classifying, organising, constructing**

5.2 EXERCISES

Geographical skills key: GS1 Remembering and understanding **GS2** Describing and explaining **GS3** Comparing and contrasting **GS4** Classifying, organising, constructing **GS5** Examining, analysing, interpreting **GS6** Evaluating, predicting, proposing

5.2 Exercise 1: Check your understanding

1. **GS1** Are the following statements true or false?
   (a) The world's volcanoes are randomly scattered over the Earth's surface.
   (b) Most of the world's volcanoes are concentrated along the edges of certain continents.
   (c) Island chains are closely linked with the location of volcanoes.
   (d) There is a weak link between the distribution of volcanoes and the location of continental plates.
   (e) Use the statements from parts a–d to write a summary paragraph, remembering to rewrite the false statements to make them true.
2. **GS2** Explain, in your own words, the meaning of subduction when referring to plate movements.
3. **GS2** Name two locations where plates are moving apart. What is happening to the sea floor in these **places**?
4. **GS1** List one example of fold, fault, dome and plateau mountains. Where is each located?
5. **GS1** State whether the following statements are true or false.
   (a) Fold mountains are the most common type of mountain in the world.
   (b) The Sierra Nevada Range was formed by the eastern part of a fault-block tilting up.
6. **GS2** How does the shape of each of the mountains shown in this subtopic provide clues as to how they were formed? How have the effects of erosion **changed** these mountains?

5.2 Exercise 2: Apply your understanding

1. **GS5** Describe the distribution of volcanoes shown in **FIGURE 1**. What does this distribution have in common with the location of plate boundaries?
2. **GS5** Look at **FIGURE 2**. How do convection currents help explain plate tectonics?
3. **GS5** Refer to **FIGURE 1**. Name three **places** where plates are converging. What mountain ranges, if any, are located in these **places**?
4. **GS6** Draw a sketch to show what you think the world's continents will look like millions of years into the future based on the way continents move and **change**. Justify your decisions.
5. **GS2** Use **FIGURES 7** and **8** to explain the formation of dome mountains.
6. **GS6** Refer to the case study in section 5.2.3, which describes the formation of the Himalayas.
   (a) Provide an explanation for why scientists found ancient sea fossils on top of the Himalayas.
   (b) Describe how the Himalayas were formed. How long did it take for the plate carrying India to crash into Asia? Explain why these mountains are described as 'young' mountains.
   (c) Based on the movements occurring, predict what might happen to the Himalayas in the future.

Try these questions in learnON for instant, corrective feedback. Go to www.jacplus.com.au.

5.3 The world's mountains and ranges

5.3.1 Where are the world's mountains?

Mountains make up a quarter of the world's landscape. They are found on every continent and in three-quarters of all the world's countries. Only 46 countries have no mountains or high plateaus, and most of these are small island nations.

Some of the highest mountains are found beneath the sea. Some islands are actually mountain peaks emerging out of the water. Even though the world's highest peak (from sea level) is Mount Everest in the Himalayas (8850 metres high), Mauna Loa in Hawaii is actually higher when measured from its base on the ocean floor. Long chains or groups of mountains located close together are called a mountain range.

FIGURE 1 The world's main mountains and mountain ranges

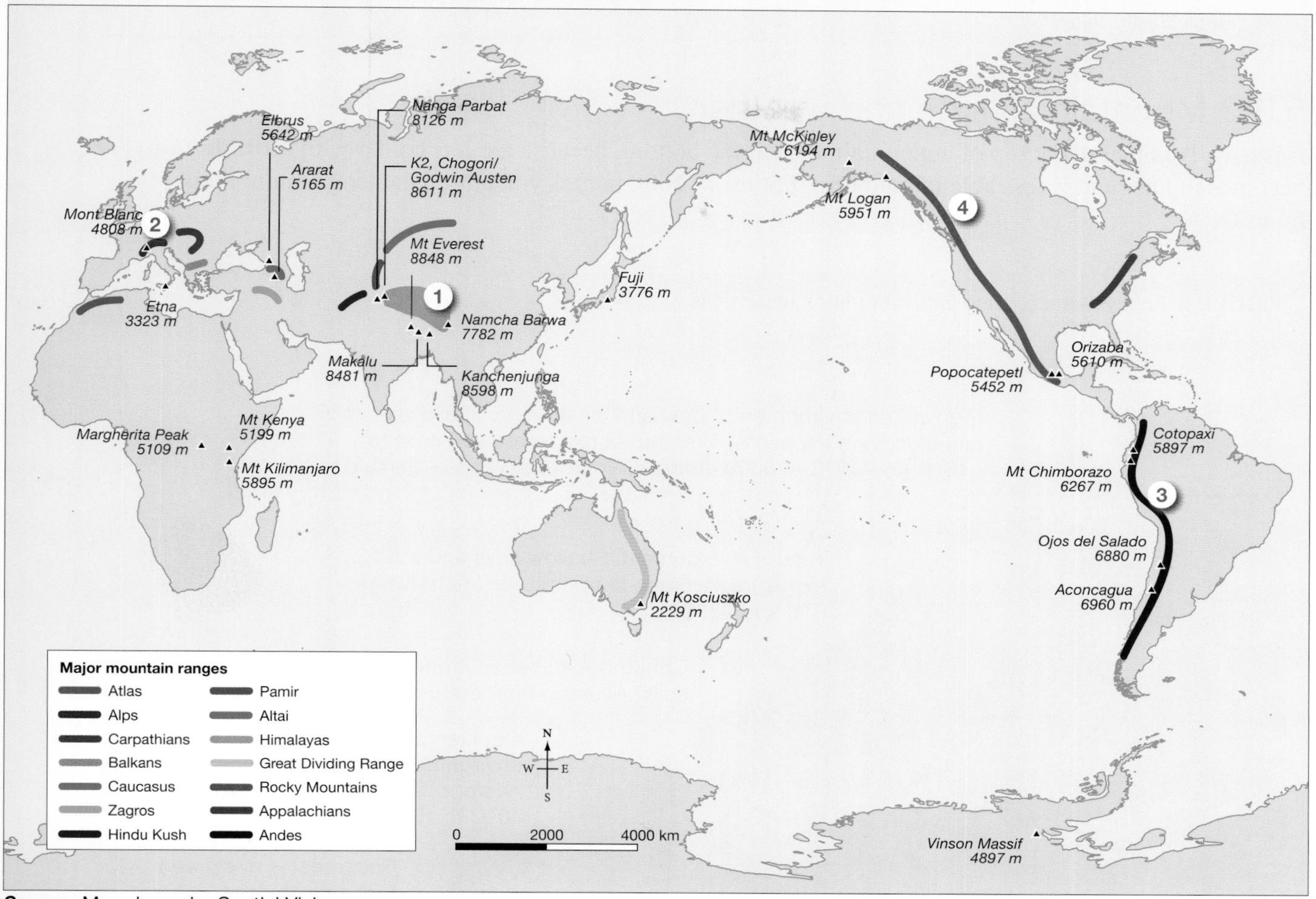

Source: Map drawn by Spatial Vision

① The Himalayas

Located in Asia, the Himalayas are the highest mountain range in the world. They extend from Bhutan and southern China in the east, through northern India, Nepal and Pakistan, and to Afghanistan in the west. The Himalayas is one of the youngest mountain ranges in the world and the name translates as 'land of snow'. The fourteen highest mountains in the world — all over 8000 metres above sea level — are all in the Himalayas.

② The Alps

The Alps, located in south central Europe, are one of the largest and highest mountain ranges in the world. They extend 1200 kilometres from Austria and Slovenia in the east, through Italy, Switzerland, Liechtenstein and Germany, to France in the west.

③ The Andes

The Andes are located in South America, extending north to south along the western coast of the continent. The Andes is the second highest mountain range in the world, with many mountains over 6000 metres. At 7200 kilometres long, it is also the longest mountain range in the world.

④ The Rocky Mountains

The Rocky Mountains in western North America extend north–south from Canada to New Mexico, a distance of around 4800 kilometres. The highest peak is Mount Elbert, in Colorado, which is 4401 metres above sea level. The other large mountain range in North America is the Appalachian Mountains, which extends 2400 kilometres from Alabama in the south to Canada in the north.

5.3.2 Mountain climate and weather

It is usually colder at the top of a mountain than at the bottom, because air gets colder with **altitude**. Air becomes thinner and is less able to hold heat. For every 1000 metres you climb, the temperature drops by 6 °C.

FIGURE 2 Ecosystems change with altitude on mountains.

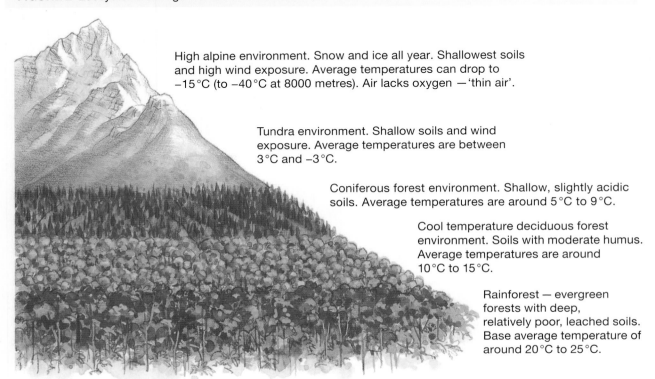

High alpine environment. Snow and ice all year. Shallowest soils and high wind exposure. Average temperatures can drop to −15 °C (to −40 °C at 8000 metres). Air lacks oxygen — 'thin air'.

Tundra environment. Shallow soils and wind exposure. Average temperatures are between 3 °C and −3 °C.

Coniferous forest environment. Shallow, slightly acidic soils. Average temperatures are around 5 °C to 9 °C.

Cool temperature deciduous forest environment. Soils with moderate humus. Average temperatures are around 10 °C to 15 °C.

Rainforest — evergreen forests with deep, relatively poor, leached soils. Base average temperature of around 20 °C to 25 °C.

5.3 INQUIRY ACTIVITY

Work in groups of 4 to 6 to investigate some of the following mountain ranges.
- Antarctica — Antarctic Peninsula, Transantarctic Mountains
- Africa — Atlas Mountains, Eastern African Highlands, Ethiopian Highlands
- Asia — Hindu Kush, Himalayas, Taurus, Elburz, Japanese Mountains
- Australia — MacDonnell Ranges, Great Dividing Range
- Europe — Pyrenees, Alps, Carpathians, Apennines, Urals, Balkan Mountains
- North America — Appalachians, Sierra Nevada, Rocky Mountains, Laurentians
- South America — Andes, Brazilian Highlands

Each student should choose a different range, and complete the following.
a. Map the location of the range in its region.
b. Describe the climate experienced throughout the range.
c. Name and provide images of a selection of plants and animals found in the range.
   Present your information in Google Maps.

Classifying, organising, constructing

5.3 EXERCISES

Geographical skills key: GS1 Remembering and understanding **GS2** Describing and explaining **GS3** Comparing and contrasting **GS4** Classifying, organising, constructing **GS5** Examining, analysing, interpreting **GS6** Evaluating, predicting, proposing

5.3 Exercise 1: Check your understanding

1. **GS1** What percentage of the Earth's surface is covered by mountains?
2. **GS1** Name the:
   (a) highest mountain range in the world
   (b) longest mountain range in the world
   (c) highest mountain in Western Europe
   (d) second-highest mountain range in North America.
3. **GS1** What name is given to long chains or groups of mountains located close together?
4. **GS1** Describe the features of a high alpine environment.
5. **GS1** What happens to oxygen in the atmosphere in high alpine environments?

5.3 Exercise 2: Apply your understanding

1. **GS5** Refer to **FIGURE 2**. How does vegetation *change* on a mountain?
2. **GS2** Refer to **FIGURE 1**. Describe how the *scale* of the world's mountains varies across the continents.
3. **GS6** Imagine you are a mountaineer, climbing to the top of Mont Blanc.
   (a) Suggest the type of clothing you need to wear for such a climb.
   (b) When you begin your climb at 1500 metres, the weather is perfect; it is sunny and clear and the temperature is 8 °C. You climb 2200 metres before you set up camp. What is the elevation? What is the temperature at this elevation? The next day the weather holds, and you climb to the summit. How far did you climb to reach the top of the mountain? What is the temperature?
4. **GS1** List the countries in which the European Alps extend.
5. **GS1** Where are the Appalachian Mountains located?

Try these questions in learnON for instant, corrective feedback. Go to www.jacplus.com.au.

5.4 SkillBuilder: Drawing simple cross-sections

online only

What are cross-sections?

A cross-section is a side-on, or cut-away view of the land, as if it had been sliced through by a knife. Cross-sections provide us with an idea of the shape of the land. We can use contour lines on topographic maps to draw a cross-section between any two points.

Cross-section of the route of Merritt's chairlift

Height above sea level (metres)

- Smooth curve
- River
- Start
- End

Select your learnON format to access:

- an overview of the skill and its application in Geography (Tell me)
- a video and a step-by-step process to explain the skill (Show me)
- an activity and interactivity for you to practise the skill (Let me do it)
- questions to consolidate your understanding of the skill.

on Resources

Video eLesson SkillBuilder: Drawing simple cross-sections (eles-1655)

Interactivity SkillBuilder: Drawing simple cross-sections (int-3151)

5.5 How people use mountains

5.5.1 Mountain people and cultures

People have moved through and lived in mountain areas for centuries. But few people live in the world's highest mountain ranges, where it can be very cold and difficult to grow food and make a living. Thousands of people visit mountains, often in remote areas, for recreation and to see the spectacular scenery, plants and animals, historic and spiritual sites, and different cultures. Mountains are also vital for global water supply.

Around 12 per cent of the world's people live in mountain regions. About half of those live in the Andes, the Himalayas and the various African mountains.

Usually, population density is very low in these areas. One reason for this is that mountains are very difficult to cross, as they are often rugged and covered with forests and wild animals. They can also be hard to climb and may have ice, snow or glaciers that make travel dangerous.

As a result of these difficulties, mountains have long provided a safe place for **indigenous peoples** and **ethnic minorities**. People live as nomads, hunters, foragers, traders, small farmers, herders, loggers and miners.

FIGURE 1 The Longshen rice terraces in China show how a mountainside can be changed to grow food

5.5.2 Mountain landscapes in the Dreamings

There are many Aboriginal and Torres Strait Islander Dreaming Stories that are linked to mountain landscapes. These teachings from the Dreamings help explain the formation and importance of each landscape and landform.

Indigenous Australian peoples are guided by Elders who know the local Dreaming Stories and customs. Dreaming Stories are passed on through the generations and explain the origin of the world around them.

The Three Sisters in the Blue Mountains

There is a story, thought to be an Indigenous Creation Story, about the formation of the Three Sisters in the Blue Mountains in New South Wales (see **FIGURE 2**). It tells of three sisters, Meehni, Wimlah and Gunnedoo, who lived in the Jamison Valley as members of the Gundungurra nations. These young women had fallen in love with three brothers from the Dharruk nation, yet tribal law forbade them to marry. The brothers were not happy with this law and so decided to use force to capture the three sisters, which caused a major battle.

As the lives of the three sisters were seriously in danger, a clever man from the Kedoombar took it upon himself to turn the three sisters into stone to protect them from any harm. He intended to reverse the spell when the battle was over, but the clever man himself was killed. As only he could reverse the spell and bring the sisters back to life, they remain in their rock formation.

FIGURE 2 The Three Sisters in the Blue Mountains

The Glasshouse Mountains

The Glasshouse Mountains located in south-east Queensland are of great historical, **cultural** and geological significance (see **FIGURE 3**). Their names — Beerwah, Tibrogargan, Coonowrin, Tunbubudla, Beerburrum, Ngungun, Tibberoowuccum and Coochin — reflect the culture of the Gubbi Gubbi people.

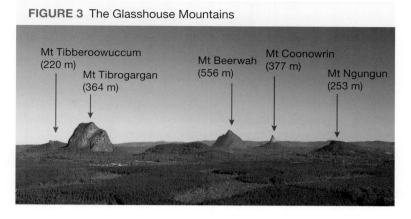

FIGURE 3 The Glasshouse Mountains

The story of these mountains goes something like this:

Tibrogargan was the father of all the nations. He and his wife, Beerwah, had many children, including Coonowrin, Tunbubudla, Miketeebumulgrai, Elimbah, Ngungun, Beerburrum and Coochin.

One day, Tibrogargan was looking out to sea when he saw the sea rising in a great swell. He became worried for Beerwah, who was pregnant. He quickly told his eldest child, Coonowrin, to take his mother to the mountains. 'I'll get the other children together and will meet you there.'

But when Tibrogargan checked to see if Coonowrin had done as he had asked, he was angered to see that he was running off alone, like a coward, and had not fetched his mother. This made Tibrogargan angry. He chased Coonowrin and hit him so hard on the head with his nulla nulla (war club) that he dislocated his neck. It has been crooked ever since.

When the flood receded, the family went back to their lands. But when the others saw Coonowrin, they teased him about his crooked neck and how he came by it, making him ashamed of his cowardice. He asked his father to forgive him, but the law would not allow this. Tibrogargan cried many tears for the shame Coonowrin had brought upon them, and his tears formed a stream that went all the way to the sea. Beerwah and all Coonowrin's brothers and sisters cried too.

Coonowrin tried to explain that he had left his mother to fend for herself because she was so big. He did not know his mother was pregnant. Tibrogargan swore he would never look at his son again, and to this day he looks at the sea and not at Coonowrin, whose head is bowed and whose tears flow into the sea. As for Beerwah — she is still pregnant.

5.5.3 Sacred and special places

Mountain landscapes often have special meaning to certain groups of people. This might be because the location includes sacred sites or religious symbols; it might also be because people want to be close to nature or to feel spiritually inspired or renewed.

Mountaineers who take great risks, climbing alone or in small groups, often find a special meaning in mountain environments. They may hold deep spiritual, cultural and aesthetic (relating to beauty) values and ideas, and these will often inspire such people to care for and protect mountain environments.

The following list gives examples of mountains that are connected to various beliefs and religions.

- Hindus and Buddhists have beliefs about Mount Kailash in the Himalayas.
- Hindus in Bali, Indonesia, have a special connection with Mount Gunung Agung.
- Tibetan Buddhists revere Chomolungma (Mount Everest).
- The landscape of Demojong in the Himalayas is sacred to Tibetan Buddhists.
- Nanda Devi in the Himalayas is a sacred site for both Sikh and Hindu communities, and is a UNESCO World Heritage site.
- Mount Fuji, in Japan, is a place of spiritual and cultural symbolism to Japanese people.
- Saint Katherine Protectorate in South Sinai, Egypt, is in an area holy to Jews, Christians and Muslims.
- Jabal La'lam is a mountain that is sacred to the people of northern Morocco.

For the indigenous groups of the north-eastern American plains, the Sioux, or Dakota as they are sometimes referred to, and the indigenous Scandinavian people, the Sami, nature was recognized as sacred. The sacred places were not man-made temples or churches, but particularly spectacular or prominent features of the natural landscape. For the Sami, these sacred places tended to be large rocks (called *sieidi*), the sides of lakes, rocky crevasses or caverns or mountaintops. These sacred mountains were somewhat isolated and had a jutting tall peak. A sacred mountain named Haldi, which rests among a group of mountains near Alta, and an 814-metre-tall conical sacred hill named Tunnsjøguden in central Norway are examples. In general, the word *saivu* is applied to sacred mountains in the south while the terms *bassi*, *ailigas* and *haldi* are used for sacred mountains by northern Sami. Similarly, mountaintops were also of spiritual importance to Sioux groups who lived in their regions; for instance, the sacred mountain Harney Peak in modern-day South Dakota.

Source: www.utexas.edu/courses/sami/diehtu/siida/religion/paralellism.htm

5.5.4 Skills to survive

It can be hard to make a living in mountain regions. People living in small, isolated mountain communities have learned to use the land and resources sustainably. Many practise shifting cultivation, migrate with grazing herds, and have terraced fields.

Some of the world's oldest rice terraces (see **FIGURE 1**) are over 2000 years old. Rice and vegetables could be grown quite densely on the terraces. This enabled people to survive in a region with very steep slopes and high altitude.

On very high ranges, below the snowline is a treeless zone of alpine pastures that can be used in summer to graze animals. Elsewhere, in the valleys and foothills, agriculture often occurs, with fruit orchards and even vineyards on some sunny slopes.

Mountains supply 60 to 80 per cent of the world's fresh water. This is due to orographic rainfall (caused by warm, moist air rising and cooling when passing over high ground, such as a mountain; as the air cools, the water vapour condenses and falls as rain). Where precipitation falls as snow, water is stored in snowfields and glaciers. When these melt, they provide water to people when they need it most.

 Resources

⊘ **Weblinks**	Dreaming stories 1
	Dreaming stories 2
	Climate change and water shortage
🛰 **Google Earth**	Glasshouse Mountains

5.5 INQUIRY ACTIVITIES

1. Use the **Dreaming stories 1** and the **Dreaming stories 2** weblinks in the Resources tab to read two Dreaming Stories. How are each of these *connected* to mountain landscapes?

   Examining, analysing, interpreting

2. From this subtopic, choose one of the mountains linked to Hindu or Buddhist beliefs. Use the internet to find out details of this *connection*. Present your information as a print or electronic brochure.

   Classifying, organising, constructing

3. Research where your water supply comes from. Which mountains, if any, are located near your water supply?

   Describing and explaining

4. Draw a consequence chart to show how and why mountains are important for water supply. Now add information to your chart about what might happen if this was reduced for some reason; for example, through climate change. Use the **Climate change and water shortage** weblink in the Resources tab to help you with this task.

   Evaluating, predicting, proposing

5.5 EXERCISES

Geographical skills key: GS1 Remembering and understanding **GS2** Describing and explaining **GS3** Comparing and contrasting **GS4** Classifying, organising, constructing **GS5** Examining, analysing, interpreting **GS6** Evaluating, predicting, proposing

5.5 Exercise 1: Check your understanding

1. **GS1** List the geographical characteristics of mountains that limit the number of people who live there.
2. **GS2** What type of work and recreation can people undertake in mountain regions? Present this information in words or in a diagram.
3. **GS2** Explain why mountains are vital for global water supply.
4. **GS2** Describe how different groups of people value mountainous *places*.
5. **GS2** Use the internet to locate the Jamison Valley in the Blue Mountains. Describe its location.

5.5 Exercise 2: Apply your understanding

1. **GS4** How has the natural mountain *environment* in **FIGURE 1** been *changed* by people? Sketch the photo and make notes to show the *changes*.
2. **GS6** Imagine you work as a park ranger in the Blue Mountains or Glasshouse Mountains. How can the Dreaming legends of the region help other people understand this *environment*?
3. **GS2** How does the Three Sisters legend help explain the formation of the Blue Mountains?
4. **GS1** Why is population density in mountain *environments* usually low?
5. **GS2** Think of a mountain you have visited or seen. Do you feel inspired by mountain *environments?* How can spiritual or religious beliefs linked to mountain landscapes help in protecting them?

Try these questions in learnON for instant, corrective feedback. Go to www.jacplus.com.au.

5.6 Earthquakes and tsunamis

5.6.1 How do earthquakes and tsunamis occur?

Earthquakes and tsunamis are frightening events and they often strike with little or no warning. An earthquake can shake the ground so violently that buildings and other structures collapse, crushing people to death. If an earthquake occurs at sea, it may cause a tsunami, which produces waves of water that move to the coast and further inland, sometimes with devastating effects.

5.6.2 Earthquakes

Earthquakes occur every day somewhere on the planet, usually on or near the boundaries of tectonic plates. The map in **FIGURE 1** of subtopic 5.2 shows a strong relationship between the location of plate boundaries and the occurrence of earthquakes. Weaknesses and cracks in the Earth's crust near these plate boundaries are called faults. An earthquake is usually a sudden movement of the layers of rock at these faults.

The point where this earthquake movement begins is called the **focus** (see **FIGURE 1**). Earthquakes can occur near the surface or up to 700 kilometres below. The shallower the focus, the more

FIGURE 1 What happens in an earthquake?

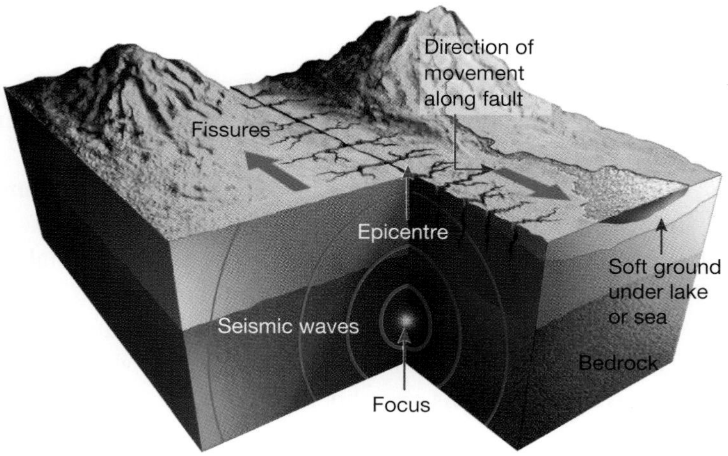

powerful the earthquake will be. Energy travels quickly from the focus point in powerful **seismic waves**, radiating out like ripples in a pond. The seismic waves decrease in strength as they travel away from the **epicentre**. The strength of an earthquake is measured on the Richter scale.

The energy released at the focus can be immense, and it travels in seismic waves through the mantle and crust of the Earth. Primary waves, or P-waves, are the first waves to arrive, and are felt as a sudden jolt. Depending on the type of rock or water in which they are moving, these waves travel at speeds of up to 30 000 kilometres an hour.

Secondary waves, or S-waves, arrive a few seconds later and travel at about half the speed of P-waves. These waves cause more sustained up-and-down movement.

Surface waves radiate out from the epicentre and arrive after the main P-waves and S-waves. These move the ground either from side-to-side, like a snake moving, or in a circular movement.

Even very strong buildings can collapse with these stresses. The energy that travels in waves across the Earth's surface can destroy buildings many kilometres away from the epicentre.

Measuring earthquakes

Earthquakes are measured according to their magnitude (size) and intensity. Magnitude is measured on the Richter scale, which shows the amount of energy released by an earthquake. The scale is open-ended as there is no upper limit to the amount of energy an earthquake might release. An increase of one in the scale is 10 times greater than the previous level. For example, energy released at the magnitude of 7.0 is 10 times greater than the energy released at 6.0.

Earthquake intensity is measured on the Modified Mercalli scale, and indicates the amount of damage caused. Intensity depends on the nature of buildings, time of day and other factors.

DISCUSS

'The strongest earthquakes result in the worst disasters.' Work in pairs or groups of three to agree, partially agree, or disagree with this statement. Use the data in this subtopic and particular examples in your response.

[Critical and Creative Thinking Capability]

5.6.3 CASE STUDY: What caused the 2015 Nepal earthquake?

On 25 April 2015, a 7.8-magnitude earthquake struck Nepal at around midday. The epicentre of this earthquake was quite shallow — only 15 kilometres below the Earth's surface. It occurred approximately 80 kilometres to the north-west of Kathmandu, Nepal's capital.

At this location, the Indian Plate to the south is subducting under the Eurasian Plate to the north (see **FIGURE 1** in subtopic 5.2). This is occurring at a rate of approximately 45 millimetres per year and is causing the uplift of the Himalayas (see the case study in section 5.2.3).

During the Nepal earthquake event, nearly 9000 people were killed and nearly 18 000 were injured (see the case study in section 5.6.8).

FIGURE 3 shows that the earthquake released a large amount of energy and caused large slips of up to four metres of the Earth's surface. There were severe aftershocks immediately after the main earthquake and the aftershocks continued for many weeks — up to 100 in total. The shaking from this earthquake was felt in China, India, Bhutan and much of western Bangladesh.

On 12 May 2015, a huge aftershock with a magnitude of 7.3 occurred near the Chinese border with Nepal (between Kathmandu and Mount Everest). More than 160 people died and more than 2500 were injured as a result of this aftershock.

FIGURE 2 The shake intensity and the tectonic plate boundary involved in the Nepal earthquake

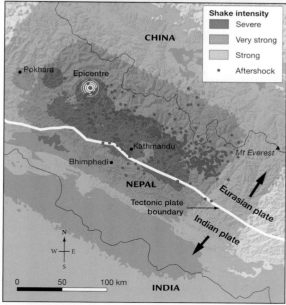

Source: USGS

FIGURE 3 Magnitudes of earthquake and aftershocks in Nepal, 2015

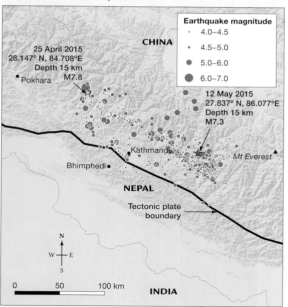

Source: USGS

5.6.4 CASE STUDY: What caused the 2011 Christchurch, New Zealand earthquake?

A 6.3–magnitude earthquake struck Christchurch, New Zealand, on 22 February 2011. The city was badly damaged, 185 people were killed and several thousand were injured. The earthquake epicentre was 10 kilometres south–east of Christchurch's central business district, and was quite shallow — only five kilometres deep, which meant the shaking was particularly destructive. The earthquake is considered to be an aftershock of an earthquake that occurred five months earlier in September 2010. Many buildings in the city had already suffered damage in the 2010 earthquake and either collapsed in the 2011 earthquake or had to be demolished afterwards.

New Zealand is located between two huge moving plates — the Australian Plate and the Pacific Plate — and it experiences thousands of earthquakes every year. Most are very small, but some have caused a lot of damage. These movements continue to shape and form New Zealand and its dramatic mountain landscapes.

FIGURE 4 Location of the Christchurch earthquake in New Zealand

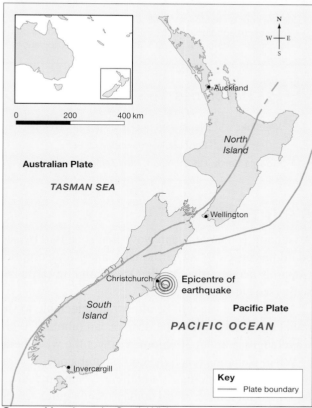

Source: Map drawn by Spatial Vision

FIGURE 5 Earthquake damage in Christchurch

Explore more with my**World**Atlas

Deepen your understanding of this topic with related case studies and questions.
• Investigate additional topics > Earthquakes and volcanoes > **Haiti earthquake**
• Investigate additional topics > Earthquakes and volcanoes > **Banda Aceh tsunami**

 Resources

 Weblinks   Nepal earthquake: before and after photos

Earthquake-vulnerable cities

Google Earth Christchurch, New Zealand

5.6.5 Tsunamis

A tsunami is a large ocean wave that is caused by sudden motion on the ocean floor. The sudden motion could be caused by an earthquake, a volcanic eruption or an underwater **landslide**. About 90 per cent of tsunamis occur in the Pacific Ocean, and most are caused by earthquakes that are over 6.0 on the Richter scale (see **FIGURE 7**).

A tsunami at sea will be almost undetectable to ships or boats. The reasons for this are that the waves travel extremely fast in the deep ocean (about 970 kilometres per hour — as fast as a large jet) and the wavelength is about 30 kilometres, yet the wave height is only one metre.

When tsunamis reach the continental slope, several things happen. The wave slows down and, as it does, the wave height increases and the wavelength decreases; in other words, the waves get higher and closer together. Sometimes, the sea may recede quickly, very far from shore, as though the tide has suddenly gone out. If this happens, the best course of action is to head to higher ground as quickly as possible.

A tsunami is not a single wave. There may be between five and 20 waves altogether. Sometimes the first waves are small and they become larger; at other times there is no apparent pattern. Tsunami waves will arrive at fixed periods between 10 minutes and two hours.

FIGURE 6 An earthquake and subsequent tsunami in the Indian Ocean in 2004 occurred along the boundary between tectonic plates.

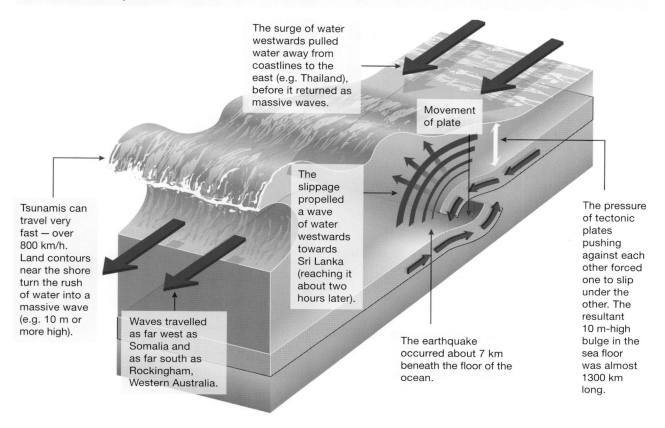

The surge of water westwards pulled water away from coastlines to the east (e.g. Thailand), before it returned as massive waves.

Movement of plate

Tsunamis can travel very fast — over 800 km/h. Land contours near the shore turn the rush of water into a massive wave (e.g. 10 m or more high).

The slippage propelled a wave of water westwards towards Sri Lanka (reaching it about two hours later).

The pressure of tectonic plates pushing against each other forced one to slip under the other. The resultant 10 m-high bulge in the sea floor was almost 1300 km long.

Waves travelled as far west as Somalia and as far south as Rockingham, Western Australia.

The earthquake occurred about 7 km beneath the floor of the ocean.

5.6.6 CASE STUDY: The Japanese tsunami, 2011

The region of Japan is seismically active because four plates meet there: the Eurasian, Philippine, Pacific and Okhotsk. Many landforms in this region are influenced by the collision of oceanic plates. Chains of volcanic islands called island arcs are formed, and an ocean trench is located parallel to the island arc (see **FIGURE 1** in subtopic 5.2).

On 11 March 2011, an 8.9–magnitude earthquake struck near the coast of Japan. The earthquake was caused by movement between the Pacific Plate and the North American Plate. It occurred about 27 kilometres below the Earth's surface along the Japan Trench, where the Pacific Plate moves westwards at about eight centimetres each year. The sudden upward movement released an enormous amount of energy and caused huge displacement of the sea water, causing the tsunami. When the tsunami reached the Japanese coast, waves more than six metres high moved huge amounts of water inland. Strong aftershocks were felt for a number of days. Nearly 16 500 people were killed and 4800 were reported missing.

FIGURE 7 The location and magnitude of the earthquake that caused the Japanese tsunami

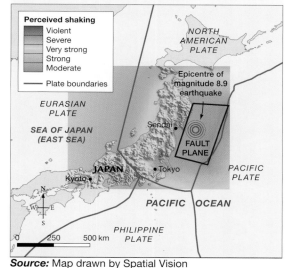

Perceived shaking
- Violent
- Severe
- Very strong
- Strong
- Moderate
—— Plate boundaries

NORTH AMERICAN PLATE

Epicentre of magnitude 8.9 earthquake

EURASIAN PLATE

SEA OF JAPAN (EAST SEA)

Sendai

FAULT PLANE

PACIFIC PLATE

JAPAN

Kyoto

Tokyo

PACIFIC OCEAN

PHILIPPINE PLATE

0   250   500 km

Source: Map drawn by Spatial Vision

FIGURE 8 This map shows the ground motion and shaking intensity from the earthquake across Japan.

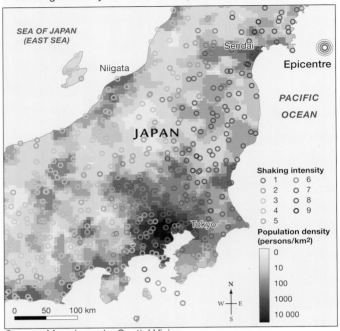

Source: Map drawn by Spatial Vision

FIGURE 9 The tsunami caused by the 8.9-magnitude earthquake in March 2011 swept over the coastline at Sukuiso and inland, carrying debris with it.

5.6.7 What are the impacts of earthquakes and tsunamis?

Earthquakes and tsunamis can have an enormous impact. The degree of impact can be affected by several factors: the size of the quake; its location; the density of the population near the epicentre; and whether there are any densely populated areas nearby. Poverty also plays a role, because it can increase a country's or region's vulnerability to such disasters. Measuring the event by the impact can be difficult. Should it be measured by the number of people killed and made homeless (social impact); the cost of recovery (economic impact); or the effect on the surroundings (environmental impact)?

Impact on people

The data in the map in **FIGURE 11** show some of the worst earthquake and tsunami disasters that have occurred. The amount of damage and death they cause does not always relate to the magnitude of the earthquake. Some smaller magnitude earthquakes can have a devastating impact. Likewise, to measure the impact of a tsunami, we have to look at its effect on people, not at the magnitude of the earthquake (or volcano) that caused it, and not at the size of the waves, which are difficult to measure.

Less developed countries often do not have the resources to prepare adequately for an earthquake. Often, many people are housed in badly constructed buildings in densely populated areas on poor land. When a disaster strikes, poorer countries often do not have the resources to act quickly and get help for relief efforts. Developed countries have strict building codes and better infrastructure to withstand disasters. They have warning systems and better communication. Usually, help is quick to arrive, with army and police personnel sent in to help with rescue efforts.

Analysis of EM-DAT (The International Disaster Database) data also shows how income levels have an impact on disaster death tolls. On average, more than three times as many people died per disaster in low-income countries (332 deaths) than in high-income nations (105 deaths). A similar pattern is evident when low- and lower-middle-income countries are grouped together and compared to high- and upper-middle-income countries. Taken together, higher-income countries experienced 56% of disasters but lost 32% of lives, while lower-income countries experienced 44% of disasters but suffered 68% of deaths. This demonstrates that levels of economic development, rather than exposure to hazards per se, are major determinants of mortality.

Impact on the environment

The impact of an earthquake or tsunami on a human environment can be catastrophic. It can damage and destroy entire settlements. Landslides can be triggered by earthquakes, permanently changing the landscape.

FIGURE 10 This landslide was caused by an earthquake in June 2008 in Honshu, Japan.

FIGURE 11 The 10 largest earthquakes and 10 most destructive tsunamis in recorded history

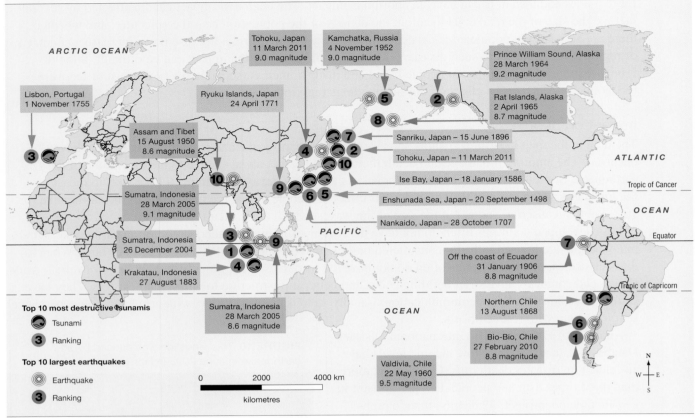

Source: Map drawn by Spatial Vision

TABLE 1 Tsunamis

No. on map and cause	Description and impact
1 9.1 earthquake	Tsunami 50 metres high, reaching 5 km inland near Meubolah. 230 000 people died. Estimated damages of US$10 billion.
2 9.0 earthquake	Tsunami waves of 10 metres swept over the east coast of Japan. 19 000 people died. Caused nuclear emergency at Fukushima Daiichi nuclear power plant. $235 billion in damage.
3 8.5 earthquake	Waves up to 30 metres high struck towns along western Portugal and southern Spain. Earthquake and tsunami killed 60 000 in Portugal, Morocco and Spain.
4 Volcano	Tsunami linked to the explosion of the Krakatau volcano. Waves as high as 37 metres demolished the towns of Anjer and Merak. Killed 40 000 people, with 2000 deaths caused by the volcanic eruptions rather than the tsunami.
5 8.3 earthquake	Homes were flooded and swept away; 31 000 people killed.
6 8.4 earthquake	Waves up to 25 metres high struck the Pacific coasts of Kyushyu, Shikoku and Honshin. Nearly 30 000 buildings were damaged in the affected regions and about 30 000 people were killed.

(continued)

TABLE 1 Tsunamis *(continued)*

No. on map and cause	Description and impact
7 7.6 earthquake (estimated)	Tsunami was reported to have reached a height of 38.2 metres, causing damage to more than 11 000 homes and killing around 22 000 people. Reports were also found of a corresponding tsunami hitting the east coast of China, killing around 4000 people and doing extensive damage to local crops.
8 Two 8.5 earthquakes	Waves up to 21 metres high affected the entire Pacific Rim for two or three days. Tsunami registered by six tide gauges as far away as Sydney, Australia. 25 000 deaths and estimated damages of US$300 million were caused along the Peru–Chile coast.
9 7.4 earthquake	Tsunami waves around 11–15 metres high destroyed 3137 homes, killing nearly 12 000 people in total.
10 8.2 earthquake (estimated)	Waves of 6 metres caused more than 8000 deaths and a large amount of damage to a number of towns.

TABLE 2 Earthquakes

No. on map and magnitude of earthquake	Description and impact
1 9.5	Killed 1655 people, injured 3000 and displaced two million. Caused US$550 million in damage. Two days later, Puyehue volcano erupted, sending ash and steam into the atmosphere for several weeks.
2 9.2	Resulting tsunami killed 128 people and caused US$311 million in damage.
3 9.1	Killed 227 900 people, displaced 1.7 million in South Asia and East Africa. On 28 December, a mud volcano began erupting near Baratang, Andamar Islands.
4 9.0	Earthquake caused tsunami that killed 19 000 people and injured 6000. Caused US$ tens of billions in damage. Economic impacts huge, especially with the shutting down of a nuclear reactor.
5 9.0	Generated a tsunami that caused damage of US$1 million in Hawaiian Islands. Some waves over 9 metres high at Kaena Point, Oahu. None killed.
6 8.8	Killed at least 521 people, with 56 missing and 12 000 injured. More than 800 000 people displaced, with a total of 1.8 million people affected across Chile, where damage was estimated at US$30 billion.
7 8.8	Earthquake caused tsunami that was reported to have killed between 500 and 1500 people in Ecuador and Colombia.
8 8.7	Generated a tsunami about 10 metres high that caused damage on Shemya Island, plus US$10 000 in property damage from flooding on Amchitka Island. No deaths or injuries reported.
9 8.6	Killed 1313 people, with more than 400 people as far away as Sri Lanka injured by the tsunami.
10 8.6	This inland earthquake caused widespread damage to buildings as well as large landslides. 780 people were killed in eastern Tibet.

Liquefaction

Liquefaction occurs when soil suddenly loses strength and, mixed with groundwater, behaves like a liquid. This usually occurs as a result of ground shaking during a large earthquake. The types of soils that can liquefy include loose sands and silts that are below the water table, so all the space between the grains is filled with water. Dry soils above the water table will not liquefy.

Once a soil liquefies, it cannot support the weight of the dry soil, roads, concrete floors and buildings above it. The liquefied soil comes to the surface through cracks, and widens them.

FIGURE 12 Cars swallowed by liquefied soil on a road in Christchurch, New Zealand, 2011

Source: © Photography by Mark Lincoln

5.6.8 CASE STUDY: Impact of the Nepal earthquake, 2015

Nearly 9000 people were killed during the 25 April and 12 May earthquakes in Nepal with more than 17 800 injured. Nearly 400 people are still missing. In addition, more than 500 000 houses were destroyed and nearly 270 000 were damaged. Nepal's historic Dharahara Tower collapsed, killing 180 people, and an avalanche at the Mount Everest Base Camp killed 21 people and injured 120. A huge avalanche also occurred in the Langtang Valley, where all the homes were destroyed and 250 people were reported missing. Hundreds of thousands of people were made homeless after buildings were destroyed or had become dangerous as a result of damage.

A further 78 people were killed and more than 500 injured in India. In China 25 people were killed and more than 380 people injured, along with 2500 homes destroyed and 24 700 damaged. The earthquake occurred during working hours so many people were outdoors. Had it occurred at night, with more people at home, the number of dead and injured would have been higher.

The economic costs are also huge — it is estimated that damage costs are between US$4–5 billion as a result of the earthquake and aftershocks. This is disastrous for a very poor country like Nepal.

For some time now, the region around Kathmandu has been known as one of the most dangerous places in the world, in terms of earthquake risk. Apart from earthquakes, other geophysical hazards that occur in Nepal include landslides, avalanches and flash flooding. In addition to its location, Nepal is extremely vulnerable because of its poverty. This means Nepal has poor building standards (many of the buildings were quickly reduced to rubble and dust) and inadequate public health and community systems to support its people in times of crisis. Without this support, clean water, safe food and effective disposal of sewerage cannot be guaranteed. There is also no adequate hospital and first aid response when disasters strike.

FIGURE 13 (a) and (b) Before and after images of Dunbar Square

(a)

(b)

DISCUSS

Earthquake engineers often say earthquakes don't kill people, collapsing buildings do. Discuss this statement in relation to poor and rich countries. What role should people in rich countries play in helping those in poor countries at risk of these events?

[Ethical Capability]

FIGURE 14 Compared to some richer countries, such as Japan and the United States, very few buildings in Nepal are earthquake-proof.

 Resources

 Interactivity Anatomy of a tsunami (int-3111)

 Weblinks P- and S-waves

World's biggest tsunami

Liquefaction

Sendai tsunami

5.6 INQUIRY ACTIVITIES

1. Use the **P- and S-waves** weblink in the Resources tab. What is the difference between the waves? How fast do they travel? How is damage caused by the waves? **Describing and explaining**

2. Use an atlas or Google Earth to find the location of Lituya Bay. Draw a map to show the location. Use the **World's biggest tsunami** weblink in the Resources tab to listen to eyewitness accounts of the event. How does this help give you a sense of the *scale* of this event? **Examining, analysing, interpreting**

3. Use the **Liquefaction** weblink in the Resources tab to view a video of liquefaction occurring. Then, write a paragraph describing what liquefaction is and why it occurs. **Describing and explaining**

4. Use the **Sendai tsunami** weblink in your Resources tab to look at satellite images showing areas before and after the 2011 Japanese tsunami. Choose two locations to draw sketches of before and after, and annotate your sketches to record the *changes* that have taken place. **Comparing and contrasting**

5. Use the **Nepal earthquake: before and after photos** weblink in the Resources panel and look at more before and after images. Choose one of the before/after images and sketch the after image, providing annotations which show the impact on people and/or the *environment*. **Comparing and contrasting**

6. Use the **Earthquake-vulnerable cities** weblink in the Resources tab to read more about cities that are most at risk from earthquakes. Use an atlas to locate these cities. Where are they located in relation to plate boundaries and, in particular, to the Pacific Ring of Fire? **Examining, analysing, interpreting**

5.6 EXERCISES

Geographical skills key: GS1 Remembering and understanding **GS2** Describing and explaining **GS3** Comparing and contrasting **GS4** Classifying, organising, constructing **GS5** Examining, analysing, interpreting **GS6** Evaluating, predicting, proposing

5.6 Exercise 1: Check your understanding

1. **GS1** What are the focus and epicentre of an earthquake?
2. **GS1** How does an earthquake occur?
3. **GS1** What does the Richter scale measure? How much more powerful is the magnitude of an earthquake at 7.0 than at 5.0?
4. **GS1** List the factors that combine to cause an earthquake or tsunami to turn into a disaster.
5. **GS1** List the different impacts that earthquakes and tsunamis can have.
6. **GS2** Geophysicists and other experts have warned for decades that Nepal was vulnerable to a deadly earthquake. Why was Nepal not better prepared for this event?
7. **GS2** How does poverty in Nepal increase vulnerability to disasters?
8. **GS5** What is the relationship between income and disaster risk? Why is the risk of earthquakes and tsunamis higher in poor countries?

5.6 Exercise 2: Apply your understanding

1. **GS5** Conduct some research to see how Japan has recovered from the 2011 tsunami and then how Nepal has recovered from the 2015 earthquake. How can you account for any differences in the recovery process?
2. **GS6** Study **FIGURE 11** in this subtopic and **FIGURE 1** in subtopic 5.2.
    (a) Describe the *interconnection* between the distribution of earthquakes and tsunamis and the distribution of tectonic plates.
    (b) Why might Japan experience so many destructive earthquakes and tsunamis?
3. **GS2** Study **FIGURE 6**. Use your own words to explain how a tsunami occurs.
4. **GS2** Study **FIGURE 2**.
    (a) In which direction is the Indian Plate moving? Is it moving under or over the Eurasian Plate?
    (b) Describe the location of the highest intensity shaking. How close was it to the epicentre? To the tectonic plate boundary?
5. **GS1** Study **FIGURE 3**. Are the following statements true or false? If they are false, rewrite them to make them true.
    (a) The earthquake and aftershocks were between 4.0 and 6.0 in magnitude.
    (b) The furthest earthquake and aftershocks were 100 kilometres apart.
    (c) The earthquake on 12 May was the same intensity as the earthquake on 25 April.
    (d) Most of the aftershocks were felt to the east of the main earthquake on 25 April.
6. **GS5** How does the earthquake event in Nepal support the idea that the Himalayas are a young mountain range that is still forming?
7. **GS2** Study **FIGURE 7**. Describe where the most violent shaking occurred as a result of the earthquake. How many plates meet in this region? What impact does this have?
8. **GS5** Study **FIGURE 8**.
    (a) Where in Japan was the greatest intensity felt?
    (b) What is the population density for Sendai, Tokyo and Niigata? How would this increase the impact of the earthquake?
9. **GS6** Study the photo of the Japanese tsunami in **FIGURE 9**.
    (a) Imagine you are a radio news reporter. Describe what you see and what might be happening to people in the area.
    (b) Imagine you were a Sendai resident. Describe what you would have done to take care of yourself.
10. **GS6** Why would most Australians not know what to do if an earthquake occurred?

Try these questions in learnON for instant, corrective feedback. Go to www.jacplus.com.au.

5.7 Volcanic mountains

5.7.1 How are volcanoes formed?

A volcano is a cone-shaped hill or mountain formed when molten magma in the Earth's mantle is forced through an opening or vent in the lithosphere. Almost all active volcanoes occur at or near plate boundaries. Some occur where two plates converge, and others occur where the plates are pulling apart, or diverging (see **FIGURE 1**). There is another group of volcanoes that are formed when plates move over hotspots.

Subduction zones

Some volcanoes are formed when an oceanic plate is pulled underneath a continental plate (see subtopic 5.2). As the crust is forced down, it heats up and becomes magma. It can then rise to the Earth's surface through a magma chamber.

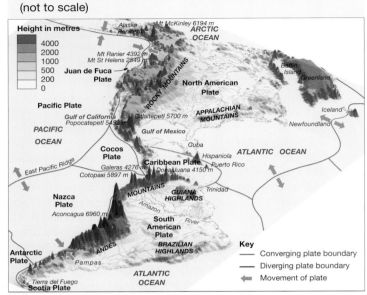

FIGURE 1 Landforms of North, Central and South America (not to scale)

Source: Map drawn by MAPgraphics Pty Ltd, Brisbane

Volcanoes in rift zones

The longest mountain range in the world is underwater between the African and American continents, and is 56 000 kilometres long. It is called the Mid-Atlantic Ridge, and it is made up of many volcanic mountains. The volcanoes are formed where two plates move away from each other in **rift zones**. The molten lava rises to the surface in the space between the plates, and the largest volcanoes appear above the water as islands. Examples of rift islands are Iceland, the Azores, Ascension Island, Gough Island and Bouvet Island. The rifting, or spreading apart, can occur on land or on the seabed.

The rifting of Iceland

The Mid-Atlantic Ridge passes through Iceland, where the island is splitting in two different areas (see **FIGURE 2**). This can be seen where Iceland's volcanoes are located, at the point where the North American Plate is drifting to the west and the Eurasian Plate is drifting to the east (see **FIGURE 3**). New crust is being formed in a rift below the sea, and eventually water from the Atlantic Ocean will fill the widening and deepening gaps between the separated parcels of land.

FIGURE 2 Rifting in Iceland

Source: Map drawn by MAPgraphics Pty Ltd, Brisbane

The Great Rift Valley, Africa

The Great Rift Valley is in Africa (**FIGURE 4**). It is about 5000 kilometres long, and stretches from Syria in the north to Mozambique in the south. The valley varies in width from 30 kilometres at its narrowest point to 100 kilometres at its widest. In some places it is a few hundred metres deep; in others it can be a few thousand metres deep.

The Great Rift Valley was created through separation that began 35 million years ago, when the African and Arabian plates began pulling apart in the northern region. About 15 million years ago, East Africa began to separate from the rest of Africa along the East African Rift. The volcanic activity in this region has produced many volcanic mountains, such as Mount Kilimanjaro, Mount Kenya and Mount Elgon.

As these rifts continue to grow, new ocean waters will flow into the valleys, separating the landmasses.

FIGURE 3 A chain of volcanoes in Iceland

FIGURE 4 The Great Rift Valley, Africa

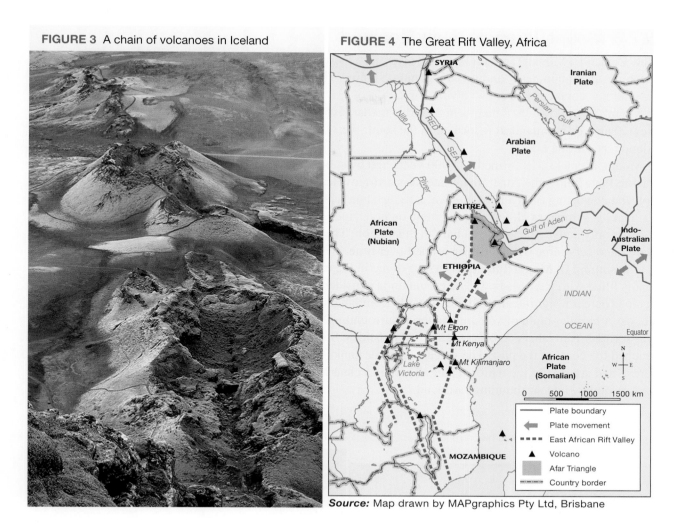

Source: Map drawn by MAPgraphics Pty Ltd, Brisbane

5.7.2 Volcano hotspots

Although most volcanoes are formed on plate boundaries, some are located in the middle of plates, a long way from plate boundaries. These volcanoes have formed above a hotspot — a single plume of rising mantle. Volcanoes form as the plates slowly move over the hotspot and, over time, a chain of volcanoes can form. Hotspots are found in the ocean and on continents. Examples include the Hawaiian Islands and many of Australia's extinct volcanoes. In Hawaii, the location of the volcanoes gives a clue to the direction and speed of the plate movement.

on Resources

🔗 **Weblink**   Hawaii's hotspot

🦫 **Google Earth**   Iceland

Great Rift Valley

─Explore more with my**World**Atlas─

Deepen your understanding of this topic with related case studies and questions.
- Investigate additional topics > Earthquakes and volcanoes > **Hawaii's hot spot**

5.7.3 Mount Taranaki, New Zealand

New Zealand's Mount Taranaki is named after the Māori terms *tara* meaning 'mountain peak' and *ngaki* meaning 'shining' (because the mountain is covered with snow in winter).

Mount Taranaki is 2518 metres high and is the largest volcano on New Zealand's mainland. It is located in the south-west of the North Island (see **FIGURE 5**).

Mount Taranaki was formed 135 000 years ago by subduction of the Pacific Plate below the Australian Plate. It is a stratovolcano — a conical volcano consisting of layers of pumice, lava, ash and tephra. Mount Taranaki is symmetrical, looking the same on both sides of a central point. It is the only active volcano in a chain in this region. The other volcanoes were once very large but have been eroded over time.

The summit of Mount Taranaki is a lava dome in the middle of a crater that is filled with ice and snow. The mountain is considered likely to erupt again. There are significant potential hazards from lahars, avalanches and floods. A circular plain of volcanic material surrounding the mountain was formed from lahars (see **FIGURE 9**) and landslides. Some of these flows reached the coast in the past. The volcano's lower flanks are covered in forest, and are part of the national park. There is a clear line between the park boundary and surrounding farmland.

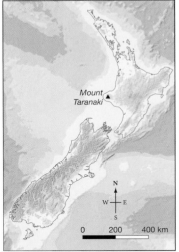

FIGURE 5 Location of Mount Taranaki on the North Island of New Zealand

Source: Map drawn by Spatial Vision

FIGURE 6 Mount Taranaki has a near-perfect conical shape.

FIGURE 7 Aerial photo of Mount Taranaki

 Resources

 Weblink   Mt Taranaki Live

  Digital document Topographic map of Mount Taranaki (doc-32264)

FIGURE 8 Topographic map of Mount Taranaki

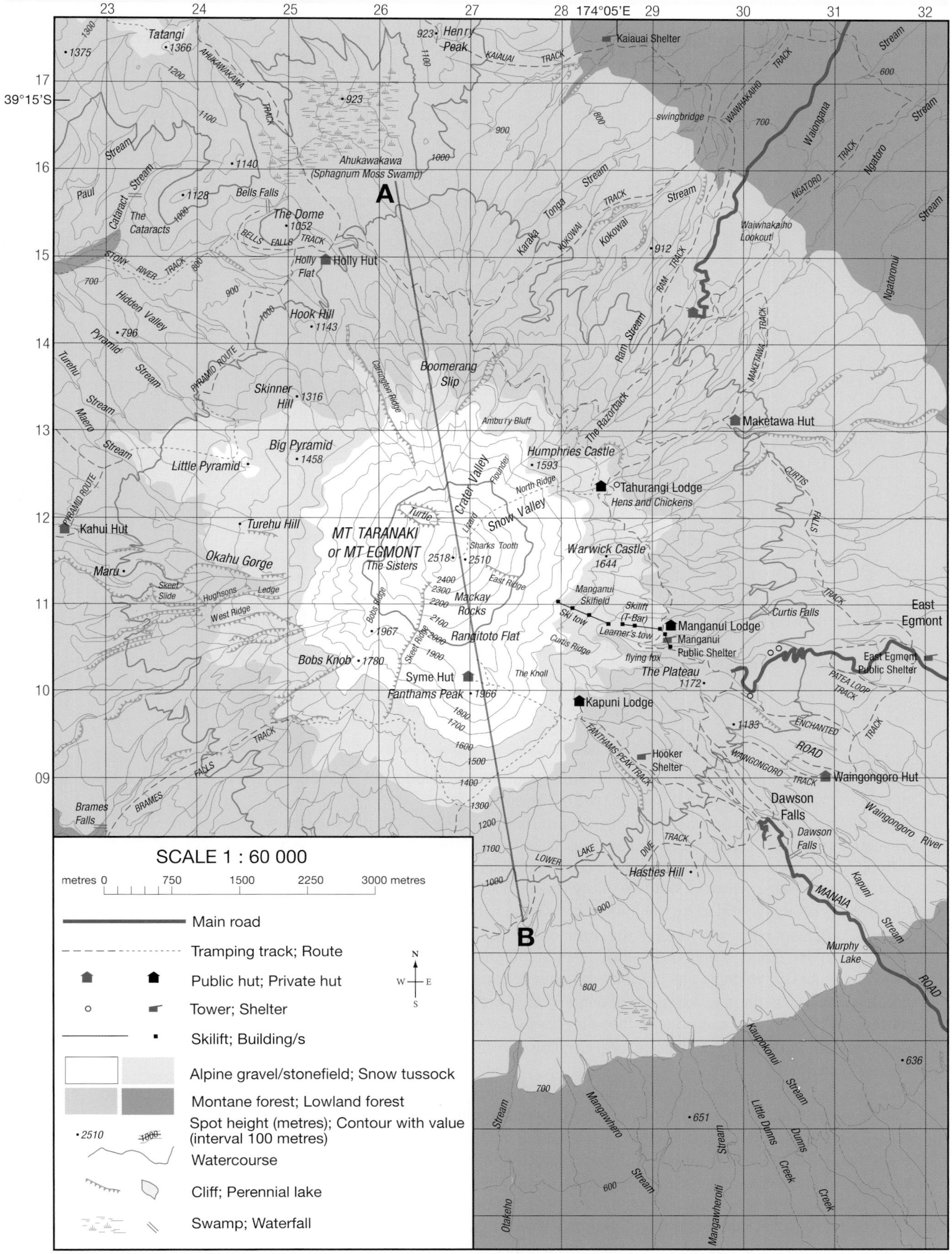

Source: Topographic Map 273-09 Egmont. Crown Copyright Reserved. Map drawn by MAPgraphics Pty Ltd, Brisbane

5.7.4 Volcanic eruptions

Volcanic mountains are formed when magma pushes its way to the Earth's surface and then erupts as lava, ash, rocks and volcanic gases. These erupting materials build up around the vent through which they erupted.

A volcanic eruption can be slow or spectacular, and can result in a number of different displays (see **FIGURE 9**).

FIGURE 9 The anatomy of a volcano

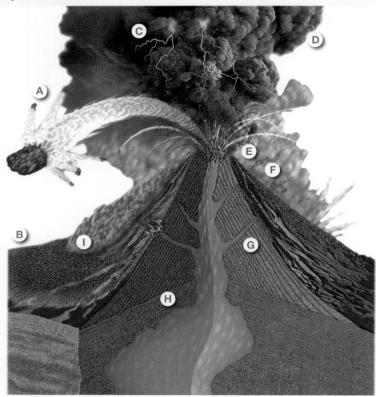

(A) A fragment of lava greater than 64 millimetres in diameter is called a volcanic bomb. They are often solid pieces of lava from past eruptions that formed part of the cone.

(B) A pyroclastic flow is a superheated avalanche of rock, ash and lava that rushes down the mountain with devastating effects. The flow can travel at up to 240 kilometres per hour and reach temperatures of 800 °C. When Mount Pelée erupted in 1902, on the island of Martinique in the Caribbean, a pyroclastic flow covered the town of Saint-Pierre, killing all but two of the town's 30 000 inhabitants.

(C) Lightning is often generated by the friction of swirling ash particles.

(D) As rock is pulverised by the force of the eruption, it becomes very fine ash, and is carried by wind away from the crater as an ash cloud. Volcanic ash may blanket the ground to a depth of many metres. In the eruption of Mount Vesuvius in 79 CE, volcanic ash completely covered two large towns: Pompeii and Herculaneum.

(E) A volcanic cone is made up of layers of ash and lava from previous eruptions. If the volcano has not erupted for thousands of years (i.e. is dormant), these layers will be eroded away.

(F) Lava may be either runny or viscous, and can flow for many kilometres before it solidifies, thereby building up the Earth's surface.

(G) Pressure may force magma through a branch pipe or side vent. In the eruption of Mount St Helens, United States, in the 1980s, the side of the mountain collapsed and the side vent became the main vent.

(H) Where two plates move apart, molten rock from the mantle flows upward into a magma chamber. More rock is melted and erupts violently upwards. Magma is generally within the temperature range of 700 °C to 1300 °C.

(I) When pyroclastic flows melt snow and ice, and mix with rocks and stones, a very wet mixture called a lahar can form. Lahars can flow quickly down the sides of volcanoes and cause much damage. One lahar that formed in 1985 on Nevado del Ruiz volcano in Colombia, travelled at up to 50 kilometres per hour and was up to 40 metres high in some places. A wall of mud, water and debris travelled 73 kilometres to the town of Armero, devastating it. More than 23 000 people died that night and 5000 homes were destroyed.

5.7.5 Volcanic shapes

Volcanoes come in a variety of shapes and sizes, forming different landforms. There are four main types and each depends on:

- the type of lava that erupts
- the amount and type of ash that erupts
- the combination of lava and ash.

Lava that is rich in silica (a mineral present in sand and quartz) is highly viscous and is thick and slow moving. If the lava is low in silica, it tends to be very runny and may flow for many kilometres before it cools and hardens to become rock. Volcanoes that erupt runny lava tend to have broad, flat sides (shield volcanoes). Those that erupt thick, treacle-like lava tend to have much steeper sides (dome volcanoes).

Heavy ash material, like volcanic bombs, settles close to the crater while lighter ash is carried further away. Volcanoes that are built up through falls of ash are steep-sided cinder cones.

The most common type of volcano is one built up of both ash and lava; this is called a composite volcano.

FIGURE 10 Four volcanic landforms

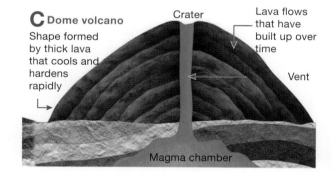

A Shield volcano

Shape formed by thin lava that spreads widely before cooling and hardening

Vent

Crater

Lava flows that have built up over time

Magma chamber

B Cinder cone

Crater

Vent

Layers made up of heavier rocks near the vent, and cinders and ash towards the edges

Magma chamber

C Dome volcano

Shape formed by thick lava that cools and hardens rapidly

Crater

Lava flows that have built up over time

Vent

Magma chamber

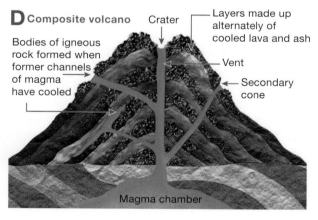

D Composite volcano

Crater

Bodies of igneous rock formed when former channels of magma have cooled

Layers made up alternately of cooled lava and ash

Vent

Secondary cone

Magma chamber

┌─ Explore more with my**World**Atlas ─────────────────────────────

Deepen your understanding of this topic with related case studies and questions.
- Investigating Australian Curriculum topics > Year 8: Landforms and landscapes > **Lahars**

5.7 INQUIRY ACTIVITIES

1. Go to the **Hawaii's hotspot** weblink in the Resources tab and explain how hotspot volcanoes form.
   Describing and explaining

2. Use an atlas to find the Cotopaxi volcano. In which country is it located? How high is it?
   Describing and explaining

3. Use an atlas or Google Earth to locate the islands on the Mid-Atlantic Ridge. Give the latitude and longitude for three locations. Describe the *interconnection* between the location of the ridge and the location of islands and volcanoes.
   Describing and explaining

4. Use the **Mt Taranaki Live** weblink in the Resources tab to view Mount Taranaki.
   Examining, analysing, interpreting

5. Use the internet to find pictures of volcanic landforms and materials. These include crater lakes, geysers, calderas, fields of ash deposits, volcanic plugs, lava tubes, hummocks and pumice. You could also find pictures of the two types of lava: a'a and pahoehoe. Use your pictures to put together a field guide to volcanic landforms. Each page should contain a picture of the landform, a brief description and a *place* where it could be found — sometimes they are tourist attractions.
   Classifying, organising, constructing

5.7 EXERCISES

Geographical skills key: GS1 Remembering and understanding **GS2** Describing and explaining **GS3** Comparing and contrasting **GS4** Classifying, organising, constructing **GS5** Examining, analysing, interpreting **GS6** Evaluating, predicting, proposing

5.7 Exercise 1: Check your understanding

1. **GS2** Refer to an atlas map of Africa and look at the shape of the island of Madagascar. Try to imagine fitting this island back into the mainland. Using plate tectonic terms, write a paragraph to describe how Madagascar's location has *changed* over time.
2. **GS2** Describe the *changes* occurring that are causing volcanoes to form in:
   (a) the Great Rift Valley
   (b) Iceland.
3. **GS2** How is the *scale* of the *changes* happening in Iceland different from the *scale* of *change* happening in the Great Rift Valley?
4. **GS2** Explain how the different shapes of volcanoes shown in **FIGURE 10** are the results of different materials being ejected.
5. **GS1** Where is Mount Taranaki located?
6. **GS1** What is a stratovolcano?
7. **GS2** Mount Taranaki receives between 3200 millimetres and 6400 millimetres of rainfall each year. How would this contribute to the shape of this landform?

5.7 Exercise 2: Apply your understanding

1. **GS2** Refer to **FIGURES 2** and **3**. Explain why a chain of volcanoes, like the one in the photograph, forms in Iceland. What is happening to the plates?
2. **GS6** Draw what you imagine Iceland will look like many thousands of years in the future after further rifting. Provide new names for each of the smaller islands. In which direction, and towards which continent, will each island drift? Describe key *changes*.
3. **GS6** Draw a series of sketches to show what you predict will happen to the African landmass as the Great Rift Valley continues to rift. Include a map of Africa showing the *change* in shape that might occur. You need to annotate your sketches to justify the predictions you have made.
4. **GS5** Refer to **FIGURE 8**.
   (a) What is the grid reference for the spot height of Mount Taranaki?
   (b) Calculate the number of private huts and public huts.
   (c) Name the ski field.
   (d) How many ski tows and lifts are there at the ski field? Calculate the length of each.
   (e) Name and give the grid reference of a lodge in which skiers could stay.
   (f) Name the other two lodges on the map.
   (g) Bushwalking is a popular activity. How many huts are on the map?

5. **GS2** Describe evidence from the aerial photo in **FIGURE 7** that the national park has protected forests around the volcano. (See the 'Interpreting an aerial photo' SkillBuilder in subtopic 5.8.)
6. **GS6** Use **FIGURES 6, 7** and **8** to describe where you think lava would flow if Mount Taranaki erupted. Describe the potential *changes* to the human and natural *environment*.
7. **GS6** Refer to **FIGURE 9**.
   (a) Describe, in detail, the *changes* to the *environment* that volcanic eruptions can cause.
   (b) Which *changes* would impact on a small *scale* and which would impact on a larger *scale*?

Try these questions in learnON for instant, corrective feedback. Go to www.jacplus.com.au.

5.8 SkillBuilder: Interpreting an aerial photo

What are aerial photos?

Aerial photographs are those that are taken from above the Earth from an aircraft.

Oblique aerial photos are those taken from an angle from an aircraft.

Vertical aerial photos are taken from directly above; that is, looking straight down onto objects.

Aerial photos can reveal details that are not recorded on maps. It is easy to see landforms with distinct shapes, different landscapes, land uses, specific places and spatial patterns of the environment.

Select your learnON format to access:

- an overview of the skill and its application in Geography (Tell me)
- a video and a step-by-step process to explain the skill (Show me)
- an activity and interactivity for you to practise the skill (Let me do it)
- questions to consolidate your understanding of the skill.

on Resources

Video eLesson SkillBuilder: Interpreting an aerial photo (eles-1654)

Interactivity SkillBuilder: Interpreting an aerial photo (int-3150)

5.9 How do volcanic eruptions affect people?

5.9.1 The worst volcanic eruptions

Volcanic eruptions both create and destroy landscapes. Most volcanic eruptions do not strike randomly but occur in specific areas, such as along plate boundaries. In some places there are high concentrations of people living near volcanoes.

Most of the world's active above-sea volcanoes are located near convergent plate boundaries where subduction is occurring, particularly around the Pacific basin. This is also the location of settlements across many countries. Over many years, volcanic eruptions have caused deaths and great damage.

How can the worst volcanoes be measured? Should it be based on the number of people killed or the cost of the damage and destruction? Or should it be the size of the explosion?

TABLE 1 The worst volcanoes based on number of deaths

Volcano	Location	Date	Number of deaths
Mt Tambora	Indonesia	5–10 April 1815	71 000+
Mt Pelee	West Indies	25 April–8 May 1902	30 000
Mt Krakatoa	Indonesia	26–28 August 1883	36 000+
Nevado del Ruiz	Colombia	13 November 1985	23 000
Mt Unzen	Japan	1792	12 000–15 000
Mt Vesuvius	Italy	AD 24 April 79	13 000+
Laki Volcanic System	Iceland	8 June 1783–February 1784	9350
Mt Kelud	Indonesia	1586	10 000
Mt Kelud	Indonesia	19 May 1919	5110

TABLE 2 The worst volcanoes based on economic impact

Volcano	Location	Date	Estimated loss (million US$)
Nevado del Ruiz	Colombia	1985	1000
Mount St Helens	USA	1980	860
Calbuco	Chile	2015	600
Mount Pinatubo	Philippines	1991	211
Galunggung	Indonesia	1982	160
Tungurahua	Ecuador	2006	150
Gamalama	Indonesia	1983	149
El Chichon	Mexico	1982	117
Rabaul	Papua New Guinea	1994	110
Puyehue-Cordon Caulle	Chile	2011	104

Source: EM-DAT International Disaster Database, January 2016 data

Explore more with myWorldAtlas

Deepen your understanding of this topic with related case studies and questions.
- Investigate additional topics > Earthquakes and volcanoes > **Mount Vesuvius**

5.9.2 Why do people live near volcanoes?

Geoscience Australia (a national organisation that provides geographic information to government) estimates that 180 million people in the Asia–Pacific region live within 50 kilometres of a dangerous volcano. There is also a strong relationship between the location of volcanoes and resources such as fertile soils, ore deposits and **geothermal energy**.

Fertile soils

Some of the most fertile soils on Earth have come from volcanic deposits of ash that is rich in nutrients, and from the physical breakdown of volcanic rocks over thousands to millions of years.

Fertile volcanic soils have been very important for rice growing in Japan and large areas of the Indonesian archipelago, especially on the islands of Java and Bali. There is also prime agriculture located in regions of rich soil; for example, around Naples, southern Italy, which generally has poor soils.

Another region of fertile volcanic soil is the agricultural area of the North Island of New Zealand. **Volcanic loam** in this area helps produce crops and pasture. Other regions include the western plains of the United States and the Hawaiian Islands. There is a small percentage of rich basalt soils in Australia, including the volcanic plains in Victoria, the north coast of New South Wales, the Scenic Rim of south-east Queensland, parts of Tasmania, and the Atherton Tablelands in north Queensland.

FIGURE 1 Agriculture and settlement near Mayon Volcano in the Philippines

Geothermal energy

Geothermal energy can be used in locations where there are active or dormant volcanoes still producing heat deep under the Earth's surface. High-temperature hot springs and geysers produce steam, which can be used to drive turbines and generate electricity. At lower temperatures, the hot water can be used for home heating or to develop hot or warm springs at resort spas. Over one quarter of Iceland's electricity is generated from geothermal heat, and it provides heating for more than 85 per cent of its homes. The other main countries that make use of geothermal heat are the United States (in California), Italy, New Zealand and Japan.

5.9.3 How to prepare for volcanic eruption

Can volcanic eruptions be predicted? What are the warning signs? How can the risk of death, injury and damage be reduced?

With about 500 million people living close to active volcanoes, it is important to watch for changes and try to predict an eruption, hopefully giving nearby residents time to evacuate.

FIGURE 2 Predicting volcanic eruptions

(A) Geologists study records of past eruptions by examining flow patterns of mud, lava and ash. From these patterns they can draw danger maps that pinpoint dangerous areas.

(B) Satellites monitor changes in gas emissions and in the shape of the volcano. Specialised equipment can also measure heat increases.

(C) Seismographs can detect the small earthquakes caused by rising magma. These are linked by transmitters to computers so that scientists can quickly detect changes.

(D) Sound-measuring equipment was used to accurately predict an eruption in Mexico in 2000.

(E) In 1983, an attempt was made to divert a lava flow away from the towns of Rocco and Rogalna on Mount Eina. A channel was dug and barriers erected. The lava slowed and solidified before reaching the towns.

(F) Samples of gas can be collected and analysed. An increase in the amount of sulfur dioxide (SO_2) may indicate that magma is moving upwards.

(G) A rise in the temperature of a crater lake often precedes an eruption.

(H) It has been suggested that explosives could be used to breach crater walls, sending lava away from towns. This was first tried in Hawaii in 1935.

(I) Helicopters have been used to drop concrete blocks in front of flowing lava.

(J) As magma rises and collects in the magma chambers, the cone may bulge outwards, warning of possible eruptions. Sensitive tiltmeters on the ground and on satellites can detect this bulging.

(K) Any bulging can also cause tiny cracks to appear.

(L) Buildings in areas prone to ash eruptions should have steeply sloping roofs so ash does not accumulate.

(M) In 1973, sea water was sprayed onto lava that was threatening a town in Iceland. The lava cooled quickly and solidified.

 Resources

🔗 **Weblink** Timeline

5.9 INQUIRY ACTIVITIES

1. (a) Use the **Timeline** weblink in the Resources tab to create a timeline of the worst volcanic eruptions, based on the information in **TABLES 1** and **2**. Include images from the internet.

   Classifying, organising, constructing

   (b) Study **TABLES 1** and **2**. Are there any *interconnections* (relationships) between the data (deaths and economic losses) and how rich or poor a country is? You may like to complete this as a class or group activity.

   Examining, analysing, interpreting

2. Watch the video about **Mount Vesuvius** and look at the information on this volcano in *myWorld Atlas*.
   (a) Where is Mount Vesuvius located? Which towns were destroyed by the volcano in AD 79?
   (b) How many people live in this volcano's immediate region?
   (c) Will all the monitors provide enough warning of an eruption for the people of Naples? Explain.
   (d) What do the scientists in this video predict for a future eruption?
   (e) What is the red zone? How large do the scientists think it should be?

   Examining, analysing, interpreting

5.9 EXERCISES

Geographical skills key: GS1 Remembering and understanding **GS2** Describing and explaining **GS3** Comparing and contrasting **GS4** Classifying, organising, constructing **GS5** Examining, analysing, interpreting **GS6** Evaluating, predicting, proposing

5.9 Exercise 1: Check your understanding

1. **GS1** Make a list of the advantages and disadvantages of living near a volcano.
2. **GS2** Is geothermal energy a renewable energy? Explain. How is this energy related to volcanic activity?
3. **GS3** Refer to an atlas map showing world population density, settlements and the location of volcanoes. Write two statements that describe the relationship between population density, settlements and volcano locations. How does this relate to people's risk?
4. **GS1** What data is collected to decide which volcanic events are the worst?
5. **GS1** What is volcanic loam and where is it found?

5.9 Exercise 2: Apply your understanding

1. **GS4** Draw a table like the one below, summarising the measures required for living with volcanoes.

Predicting eruptions	Preparing for eruptions	Lessening the effects of eruptions

   (a) Use it to help you classify the information in **FIGURE 2**.
   (b) Which of these measures do you think are most effective? Give three reasons for your answer.
2. **GS4** Draw a photo sketch of **FIGURE 1** and label the following: volcano; volcanic plain; lava flows; farmland; settlement.
3. **GS6** What is geothermal energy? What do you think could be some of the benefits of using this type of energy?
4. **GS2** How can seismographs be used to warn of a possible volcanic eruption?
5. **GS5** Study **FIGURE 2**. List the different techniques that have been used to try to stop the flow of lava. Which technique do you think is the most effective and why?

Try these questions in learnON for instant, corrective feedback. Go to www.jacplus.com.au.

5.10 Thinking Big research project: Earthquakes feature article

SCENARIO

Congratulations! You have been promoted to feature writer of the *Weekly Rattle*, a leading geographical magazine. Your first brief is to write a feature about the strongest earthquakes that occur in the world over a one-week period.

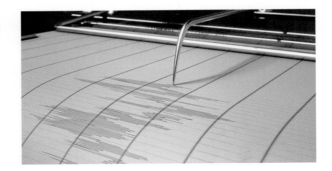

Select your learnON format to access:

- the full project scenario
- details of the project task
- resources to guide your project work
- an assessment rubric.

 Resources

 projectsPLUS Thinking Big research project: Earthquakes feature article (pro-0171)

5.11 Review

5.11.1 Key knowledge summary

Use this dot point summary to review the content covered in this topic.

5.11.2 Reflection

Reflect on your learning using the activities and resources provided.

 Resources

 eWorkbook Reflection (doc-31350)

Crossword (doc-31351)

Interactivity Mountain landscapes crossword (int-7598)

KEY TERMS

altitude height above sea level

convection current a current created when a fluid is heated, making it less dense, and causing it to rise through surrounding fluid and to sink if it is cooled; a steady source of heat can start a continuous current flow

converging plates a tectonic boundary where two plates are moving towards each other

cultural relating to the ideas, customs and social behaviour of a society

divergent plates a tectonic boundary where two plates are moving away from each other and new continental crust is forming from magma that rises to the Earth's surface between the two

epicentre the point on the Earth's surface directly above the focus of an earthquake

ethnic minority a group that has different national or cultural traditions from the main population

fault an area on the Earth's surface that has a fracture, along which the rocks have been displaced

focus the point where the sudden movement of an earthquake begins

geothermal energy energy derived from the heat in the Earth's interior

hotspot an area on the Earth's surface where the crust is quite thin, and volcanic activity can sometimes occur, even though it is not at a plate margin

indigenous peoples the descendants of those who inhabited a country or region before people of different cultures or ethnic origins colonised the area

landslide a rapid movement of rocks, soil and vegetation down a slope, sometimes caused by an earthquake or by excessive rain

liquefaction transformation of soil into a fluid, which occurs when vibrations created by an earthquake, or water pressure in a soil mass, cause the soil particles to lose contact with one another and become unstable; for this to happen, the spaces between soil particles must be saturated or near saturated

lithosphere the crust and upper mantle of the Earth

Pangaea the name given to all the landmass of the Earth before it split into Laurasia and Gondwana

rift zone a large area of the Earth in which plates of the Earth's crust are moving away from each other, forming an extensive system of fractures and faults

seismic waves waves of energy that travel through the Earth as a result of an earthquake, explosion or volcanic eruption

volcanic loam a volcanic soil composed mostly of basalt, which has developed a crumbly mixture

6 Rainforest landscapes

6.1 Overview

We can plant new trees anytime and anywhere. What makes the world's rainforests so special?

6.1.1 Introduction

What do you know about rainforest landscapes? Did you know that rainforests have the greatest biodiversity of any forest environment? They contain complex layers that support thousands of species of plants and animals. The rainforest has supplied resources to all people, including indigenous communities. People are concerned that clearing large areas of this landscape is creating negative impacts that are unsustainable. In this topic we will look at rainforest landscapes around the world, how people use them and the threats these important environments face.

On Resources

☑ **eWorkbook**   Customisable worksheets for this topic

▦ **Video eLesson** Protecting our landscapes: rainforest (eles-1627)

LEARNING SEQUENCE

6.1 Overview
6.2 Rainforest characteristics
6.3 **SkillBuilder:** Creating and describing complex overlay maps
6.4 Changing rainforest environments
6.5 **SkillBuilder:** Drawing a précis map   online only
6.6 Indigenous peoples and the rainforest
6.7 Disappearing rainforests
6.8 Social and environmental impacts of deforestation
6.9 Saving and preserving rainforests
6.10 **Thinking Big research project:** Rainforest display   online only
6.11 **Review**   online only

To access a pre-test and starter questions and receive immediate, **corrective feedback** and **sample responses** to every question, select your learnON format at www.jacplus.com.au.

6.2 Rainforest characteristics

6.2.1 Rainforests

Forests that grow in constantly wet conditions are defined as rainforest landscapes. A rainforest is an example of a biome (a community of plants and animals spread over a large natural area). Rainforests are located wherever the annual rainfall is more than 1300 millimetres and is evenly spread throughout the year. While tropical rainforests are the best known of these landscapes, there are also other types.

6.2.2 Tropical rainforests and their processes

Tropical rainforest landscapes are found where there are both high temperatures and high precipitation. The sun's rays that reach the Earth near the equator have a smaller area of the Earth and atmosphere to heat than rays reaching the Earth at higher latitudes. Therefore, it is hotter at the equator than at higher latitudes. Rainforests are also generally warmer at night, because the cloud cover and high humidity help to keep the heat in. Tropical rainforests have a hot climate right throughout the year with no summer or winter. High precipitation around the equator is mainly due to convectional rainfall and is often associated with thunderstorms. Convectional rainfall occurs when warm, moist air is heated when it moves over a hot surface on Earth. As the air is heated it expands and becomes lighter than the surrounding air. This causes it to rise. If the air continues to rise, condensation and precipitation occur. This combination of high temperatures and high precipitation influences the global distribution of the tropical rainforest landscape. Plants flourish in these rainforests, which support a huge number of plants and animals — perhaps as many as 90 per cent of all known **species**. Poison-dart frogs, birds of paradise, piranha, tarantulas, anacondas, Komodo dragons and vampire bats are all found in tropical rainforests.

Tropical rainforests that occur in the mountains, 1000 metres or more above sea level, are called montane rainforests. Other tropical rainforests are known as lowland rainforests (see **FIGURE 1**).

FIGURE 1 World rainforest types

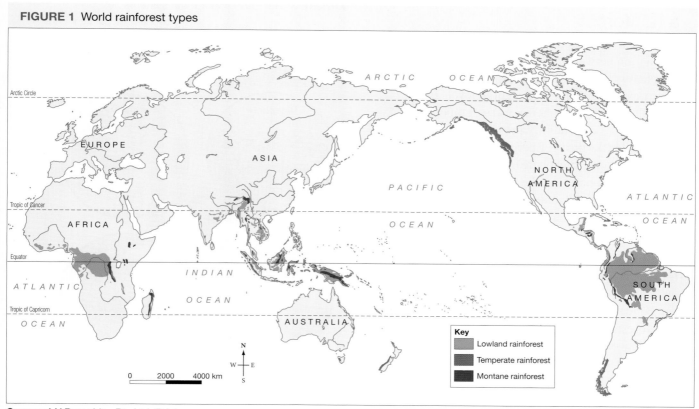

Source: MAPgraphics Pty Ltd, Brisbane

Lowland tropical rainforest

Lowland tropical rainforests form the majority of the world's tropical rainforests. They grow at elevations generally below 1000 metres. Trees in lowland forests are usually taller than those in montane forest and include a greater diversity of fruiting trees. These attract animals and birds adapted to feed on their fruits. These rainforests are far more threatened than montane forests because of their accessibility, soils that are more suitable for agriculture and more valuable hardwoods for timber. Lowland forests occur in a belt around the equator, with the largest areas in the Amazon Basin of South America, the Congo Basin of central Africa, Indonesia and New Guinea.

FIGURE 2 (a) Montane rainforest, (b) temperate rainforest and (c) lowland rainforest

Temperate rainforests

The large area of the globe between the tropics and the polar regions (areas within the Arctic and Antarctic circles) is called the **temperate** zone, and rainforests can grow there too. Temperate rainforests occur in North America, Tasmania, New Zealand and China. Giant pandas, Tasmanian devils, brown bears, cougars and wolves all call temperate rainforests home.

6.2.3 Physical processes of a rainforest

Rainforest landscapes are the result of the interaction between the Earth's four main systems or spheres. For example, the trees in a tropical rainforest (biosphere) rely on high levels of precipitation (hydrosphere), warm temperatures (atmosphere) and stability provided by soil (lithosphere) to thrive. Energy from the sun is stored by plants (biosphere). When humans or animals (biosphere) eat the plants, they acquire the energy originally captured by the plants.

6.2.4 Rainforest ecosystems

Rainforests are unique **ecosystems** consisting of four different layers — the emergent, canopy and understorey layers and the forest floor. Each layer can be identified by its distinct characteristics. Rainforests are actually a community of plants and animals working together to survive, linked in a food web (see **FIGURE 4**).

FIGURE 3 The Earth's four main systems

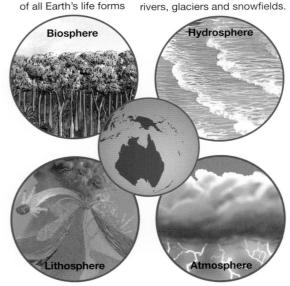

Biosphere: the collection of all Earth's life forms

Hydrosphere: 97 per cent of the Earth's water is found in salty oceans, and the remainder as vapour in the atmosphere and as liquid in groundwater, lakes, rivers, glaciers and snowfields.

Lithosphere: consists of the core, mantle and crust of the Earth

Atmosphere: contains all of the Earth's air

Emergents

These are the tallest trees, ranging in height from 30 to 50 metres. They are so named because they rise up or emerge out of the forest canopy. Huge crowns of leaves and abundant animal life thrive on plenty of available sunlight.

Canopy

This describes the array of treetops that form a barrier between the sunlight and the underlying layers. Their height can vary from 20 to 45 metres. This layer contains a distinct **microclimate** and supports a variety of plants and animals. The taller trees host special vines called lianas that intertwine the branches. Other plants called epiphytes use the tree trunks and branches as anchors in order to capture water and sunlight.

Understorey

This layer contains a mixture of smaller trees and ferns that receive only about five per cent of the sun's energy. Many animals move around in the darkness and humidity, using the vines as highways.

Forest floor

This bottom layer is dominated by a thick carpet of leaves, fallen trees and huge buttress roots that support the giant trees above. Rainforest soils give the impression of being fertile because they support an enormous number of trees and plants. However, this impression is wrong, as the soil in rainforests is generally poor. Leaves and other matter are recycled by the many organisms to create an organic **compost**. The roots of trees must 'snatch' these nutrients from the soil before heavy rains wash them away and they are lost through a process called **leaching**.

Larger animals also roam through this layer in search of food.

FIGURE 4 Layers in a tropical rainforest

Canopy

Tall emergent tree

Liana

Buttress roots

Epiphytes

Ferns

Undergrowth

Moss

 Resources

🦖 **Interactivity** Our living green dinosaurs (int-3112)

🔗 **Weblink** Rainforest layers

FIGURE 5 An example of a typical food web in an Australian rainforest

6.2 INQUIRY ACTIVITIES

1. Many rainforest animals live their whole life in the trees. Using the internet to help you, give some examples of these animals and conduct research into the habits of one animal. **Classifying, organising, constructing**
2. Use the **Rainforest layers** weblink in the Resources tab to explore the layers of the rainforest and the plants and animals that inhabit them. **Examining, analysing, interpreting**

6.2 EXERCISES

Geographical skills key: GS1 Remembering and understanding **GS2** Describing and explaining **GS3** Comparing and contrasting **GS4** Classifying, organising, constructing **GS5** Examining, analysing, interpreting **GS6** Evaluating, predicting, proposing

6.2 Exercise 1: Check your understanding

1. **GS1** What conditions do rainforest *environments* thrive in?
2. **GS2** What are the differences between montane and lowland rainforest *environments*? What causes these *changes* in rainforest type?
3. **GS2** Study **FIGURE 1**. Describe the distribution of rainforests around the world. Think about in which continents and between which latitudes they are found, the size and *scale* of them, and whether they are continuous or scattered.

4. **GS2** Imagine you are a raindrop. Recreate your journey through a rainforest, passing through each of the forest layers. Read or act out your descriptions to the rest of the class.
5. **GS1** Identify key characteristics of a tropical rainforest.
6. **GS2** Why are lowland rainforest **environments** more threatened by human activity than montane rainforests?
7. **GS1** Why are montane forests often called 'cloud forests'?
8. **GS1** How many layers are there in a rainforest **environment**?
9. **GS1** What are the tallest trees in the rainforest called?
10. **GS2** Describe how conditions in the canopy layer differ from those on the forest floor.

6.2 Exercise 2: Apply your understanding
1. **GS5** Refer to **FIGURE 1**.
   (a) Use an atlas to help you name six countries in the Asia–Pacific region that contain rainforests.
   (b) What type of rainforest **environment** is found:
       i. in north-eastern Australia
       ii. along the western coastline of Canada?
2. **GS4** Refer to **FIGURE 3**. List Earth's four spheres. Give several examples of features in each sphere.
3. **GS2** List some Earth sphere interactions from your own daily activities.
4. **GS2** Why are rainforest **environments** able to support a large range of animals and plants?
5. **GS4** Draw up and complete a table like the one below that summarises the features of a rainforest **environment**.

Layer	Height	Amount of light	Features

6. **GS2** In a rainforest, the soil below the trees is often poor and shallow, and the trees create their own nutrients. In one sentence describe how this happens, and draw a labelled sketch to illustrate the process.
7. **GS6** What **change** might you expect in the success of plant growth if the rainforest trees are removed and crops are planted instead? Why?

Try these questions in learnON for instant, corrective feedback. Go to www.jacplus.com.au.

6.3 SkillBuilder: Creating and describing complex overlay maps

What is a complex overlay map?
A complex overlay map is created when one or more maps of the same area are laid over one another to show similarities and differences between the mapped information. Traditionally, the second map is on tracing paper that is attached to the original page.

Complex overlay maps show relationships between factors — the similarities and the differences in a pattern.

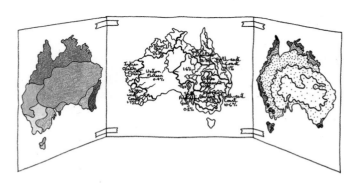

Select your learnON format to access:
- an overview of the skill and its application in Geography (Tell me)
- a video and a step-by-step process to explain the skill (Show me)
- an activity and interactivity for you to practise the skill (Let me do it)
- questions to consolidate your understanding of the skill.

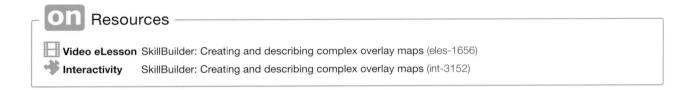
6.4 Changing rainforest environments

6.4.1 Australian rainforests

Australia's Daintree rainforest is the Earth's oldest tropical rainforest. It is estimated to be around 180 million years old and developed tens of millions years before the Amazon rainforest in Brazil. Although hard to believe now, Australia was once mostly covered in rainforest! Even areas that today are deserts were once teeming with plant and animal life similar to those in the Amazon. This is because Australia was further north than it is today. Over the past 100 million years, however, a series of events has gradually reduced the area of Australia's rainforests (see **FIGURE 1**).

FIGURE 1 The development of Australian rainforests through time

Triassic period (245–208 mya)
• Foundation of Australian rainforests was occurring as massive volcanic eruptions laid down thick layers of ash near freshwater lakes.

Early Cretaceous period (144–95 mya)
• Australia, as part of Gondwanaland, was very close to the South Pole and experienced the same conditions as present polar regions. However, the climate was temperate, supporting plant life, and the first flowering plants appeared.

Late Cretaceous period (95–66.4 mya)
• The climate was temperate and humid; species found in rainforests today began to develop. Dinosaurs were still present.

Present

260 240 220 200 180 160 140 120 100 80 60 40 20

[Dates are expressed in mya — million years ago.]

Jurassic period (208–144 mya)
• Ash deposits developed into sedimentary rock; plants such as the cycad and southern pine first appeared.

Late Oligocene/early Miocene periods (25–20 mya)
• The whole of Australia was covered by rainforest at the start of this period, but by the end of the Miocene sclerophyll forest and grasslands were starting to emerge. The rainforests would have been similar to those of today.

The gradual movement of Australia southwards as it separated from Gondwanaland and a series of **ice ages** have combined to make it a drier place (see **FIGURE 2**). Rainforests have become confined mainly to the mountains and **gorges** of the Great Dividing Range and Tasmania. These areas have higher rainfall and fewer fires.

The farming practices of European immigrants have reduced much of the remaining rainforest — in the past 200 years, more than 70 per cent of these forests have been cleared.

FIGURE 2 Difference in the location of Australia when it formed part of Gondwanaland compared with its location today

Gondwanaland   Present day

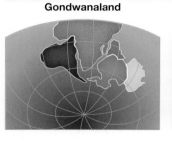 

Scientists have identified three major types of rainforest in Australia (see **FIGURE 3**). There are examples of all three in Queensland. This diversity occurs nowhere else on Earth.

FIGURE 3 Australia's rainforest areas

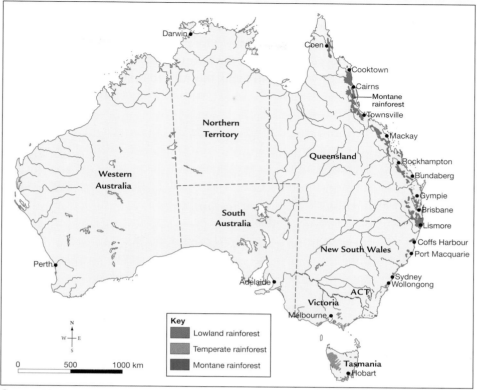

Key
- Lowland rainforest
- Temperate rainforest
- Montane rainforest

0   500   1000 km

Source: MAPgraphics Pty Ltd, Brisbane

Much of Australia's tropical rainforests are now World Heritage areas. This means they have been listed by UNESCO as being of global importance. The Wet Tropics of Queensland are a World Heritage area containing some of the oldest rainforests in the world. They have the world's highest concentration of flowering plants, and have records that show Aboriginal communities are the world's oldest indigenous rainforest culture.

The Indigenous inhabitants of the Daintree Rainforest in North Queensland are the Kuku Yalanji Aboriginal people, believed to have lived in this area for more than 9000 years by European estimates. Their culture is uniquely adapted to the rainforest environment.

For the Kuku Yalanji, the natural world is often thought of in human terms and is closely linked to the people. Any changes to the environment are seen as changes to themselves. Because of the powerful properties attributed to most story places (sites with links to the Dreamings) of the Daintree, the Kuku Yalanji regard damage and destruction to the environment as unacceptable.

The Kuku Yalanji people gather their food and medicine and many of their implements, weapons, fibres and construction material from plants in their environment. The natural patterns and cycles of the rainforest give important information about the food that is available. The plants are their calendar, marking the seasons. For example, when blue ginger (jun jun) is fruiting it is time to catch scrub turkey (diwan), and when mat grass (jilngan) is flowering it is time to collect the eggs of the scrub fowl (jarruka).

FIGURE 4 Rainforest plants were used to make goods such as these baskets that were used for storage, food collection, carrying personal possessions, and leaching poisons (from seeds) in fresh running water.

6.4.2 The Amazon rainforest

The world's largest remaining rainforest is in the Amazon Basin in South America. This truly remarkable forest is under increasing threat from forestry, mining and farming. The loss may cause severe problems worldwide. Most of us use rainforest products every day. More importantly, however, rainforests help control the world's climate and our oxygen supply. So the next time you eat chocolate, treat your asthma, play a guitar or even take a deep breath, you should thank the Amazon rainforest.

FIGURE 5 The Amazon Basin

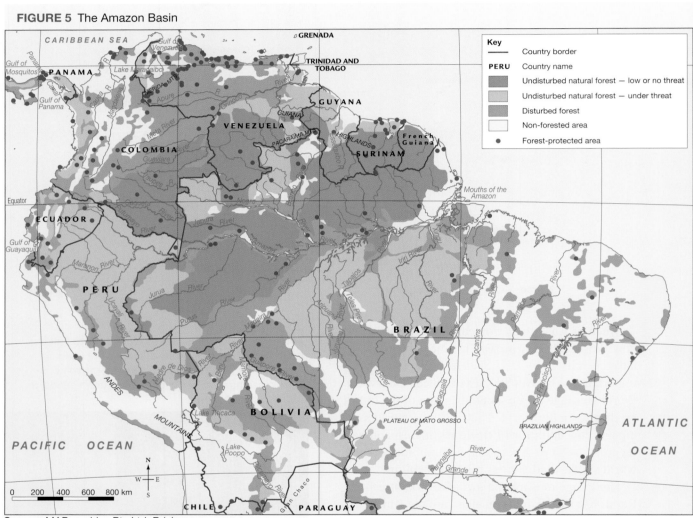

Source: MAPgraphics Pty Ltd, Brisbane

- The Amazon River and those rivers that feed into it (tributaries) contain one-fifth of the world's fresh water, and more than 2000 species of fish — more than in the Atlantic and Pacific oceans combined.
- The mouth of the Amazon River is approximately 325 kilometres wide and contains an island the size of Switzerland!
- The Amazon forest is home to more than 40 000 species of plants, 1300 bird species, 430 different mammals and 2.5 million different insects.
- Approximately 1.3 million tons of sediment is transported by the Amazon River to the sea daily.
- No bridges cross the main trunk of the Amazon River, which locals call the Ocean River.
- Since 2000, the Amazon rainforest has been facing deforestation at an average rate of 50 football fields per minute.
- The Amazon is the second longest river in the world, but it carries more water than the next six largest rivers combined.
- The Amazon River drains nearly 40 per cent of South America.
- There are official plans for 412 dams to be in operation in the Amazon River and its headwaters.
- Since 1900, more than 90 indigenous groups have disappeared in Brazil alone.

FIGURE 6 The brown waters of the Amazon show that it is carrying a lot of sediment.

FIGURE 7 Area cleared for ranching in the Amazon rainforest

FIGURE 8 Development is clearly visible within the green carpet of the Amazon rainforest between 1975 (left) and 2012 (right).

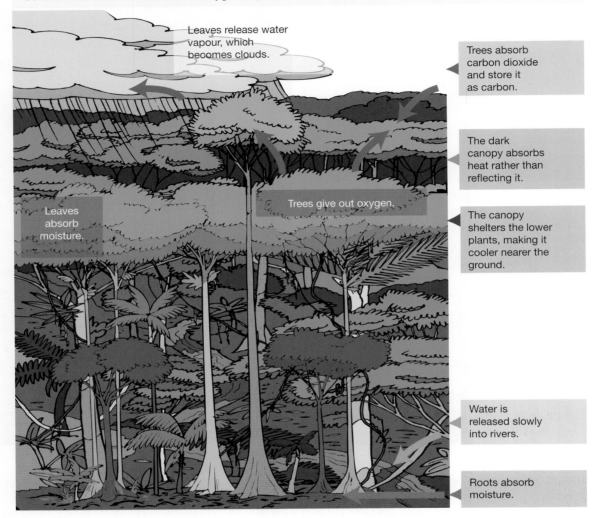

FIGURE 9 Rainforests play a vital role in controlling the world's climate and oxygen supply. Scientists believe that half of all the world's oxygen is produced by the Amazon rainforest alone.

Leaves release water vapour, which becomes clouds.

Trees absorb carbon dioxide and store it as carbon.

The dark canopy absorbs heat rather than reflecting it.

Trees give out oxygen.

The canopy shelters the lower plants, making it cooler nearer the ground.

Leaves absorb moisture.

Water is released slowly into rivers.

Roots absorb moisture.

6.4.3 Amazing rainforests

- More than 7000 modern medicines are made from rainforest plants. They can be used to treat problems from headaches to killer diseases like malaria. They are used by people who suffer from multiple sclerosis, Parkinson's disease, leukaemia, asthma, acne, arthritis, diabetes, dysentery and heart disease among many others.
- Even animals can be used to cure human diseases. Tree frogs from Australia give off a chemical that can heal sores, and a similar chemical from a South American frog is used as a powerful painkiller.
- The poisonous venom from an Amazonian snake is used to treat high blood pressure.
- Only one per cent of the known plants and animals of the rainforest have been properly analysed for their medicinal potential. Perhaps the greatest benefits to medicine and our own health, therefore, are yet to come.

FIGURE 10 Skin secretions from frogs such as the Waxy Monkey Treefrog (*Phyllomedusa bicolor*) contain powerful painkillers.

- Rainforests are home to the greatest profusion of life on the planet: at least half of all known plants and animals live in rainforests.
- At least 50 million indigenous peoples live in rainforests worldwide. From the Kuna people of Panama and the Yanomami of Brazil to the Baka people of Cameroon and the Penan of Borneo (Indonesia), these people have traditionally lived a way of life that has little impact on their forest home.
- The people who live in or near the rainforests gain much of their food from the forest. But rainforests also supply the supermarkets of the world with their bounty. Most of these fruits and nuts are now grown by farmers rather than harvested directly from the forest, but it was in the rainforests that they originated.
- Chocolate first came from cacao trees native to the Amazon rainforest. Today the cocoa in the chocolate you eat is most likely to have come from huge cacao plantations in West Africa. Similarly, brazil and cashew nuts, cinnamon, ginger, pepper, vanilla, bananas, pineapples, coconuts, paw-paws, mangoes and avocados were all originally rainforest plants. Even the gum used in chewing gum comes from a rainforest plant, as does the tree that produces rubber.
- Rainforest trees are generally hardwood trees, making them resistant to decay and attractive for building. Well-known rainforest timbers are mahogany, teak, ebony, balsa and rosewood. Rosewood is particularly interesting, as it is considered the best timber in the world for guitar making. In many tropical countries, people also collect timber as fuel for cooking or heating.

FIGURE 11 The Kamayurá people of the Brazilian rainforest live a traditional way of life.

FIGURE 12 Food products such as chocolate and chewing gum are made from ingredients that originally came from the rainforest.

 Resources

Weblinks   UNESCO Heritage

Rainforest foods

Amazon tour

6.4 INQUIRY ACTIVITIES

1. Use the **UNESCO Heritage** weblink in the Resources tab to complete the following.
   (a) On a map of Australia, locate and label Australia's World Heritage sites.
   (b) Which three sites have been added most recently?
   (c) Which two sites protect Australian rainforests?
   (d) The Wet Tropics of Queensland are particularly special because they border another World Heritage site. What is this other site?
   (e) What criteria does UNESCO use to determine whether a natural region should be placed on its list?

   Describing and explaining

2. A hotel chain has applied to the Queensland government for permission to build a resort in the Daintree. Assess this proposal from the perspectives of the developers, government, local residents, environmentalists and Kuku Yalanji people. Try to make a decision as to whether this project should be approved. This could be completed in small groups or debated as a class. **Evaluating, predicting, proposing**

3. Using a piece of tracing paper, trace the Amazon River and its tributaries. Draw a single line that joins the source of each of the tributaries. Shade the area within this line using a light blue pencil: this area is known as the **catchment**, or basin, of the river. Overlay your completed diagram on the map of the forest and comment on the *interconnection* between the river and the forest. **Classifying, organising, constructing**

4. Use the **Rainforest foods** weblink in the Resources tab to learn how the food you eat comes from the rainforest. **Examining, analysing, interpreting**

5. Use the **Amazon tour** weblink in the Resources tab to take a tour through an Amazon rainforest slideshow. **Examining, analysing, interpreting**

6. Make a list of things in your home that may come from the rainforest *environment*. Remember to look in the medicine cupboard and the pantry as well as at the furniture. Perhaps you could bring some examples to school and your class could set up a display. **Classifying, organising, constructing**

7. This subtopic lists only a few of the products we use from rainforests. List the value of these and other rainforest products under the following headings.
   (a) Valued by different cultures
   (b) Valued economically
   (c) Valued for its aesthetic value (beauty)
   (d) Other

   Classifying, organising, constructing

6.4 EXERCISES

Geographical skills key: GS1 Remembering and understanding **GS2** Describing and explaining **GS3** Comparing and contrasting **GS4** Classifying, organising, constructing **GS5** Examining, analysing, interpreting **GS6** Evaluating, predicting, proposing

6.4 Exercise 1: Check your understanding

1. **GS1** List the reasons given in this subtopic for the gradual disappearance of Australia's rainforest *environments*.
2. **GS1** What resources does the rainforest provide for the Kuku Yalanji people?
3. **GS2** How has the *scale* of Australian rainforest *environments changed* over time?
4. **GS2** Describe the location and distribution of Australia's remaining rainforest *environments*. What factors have contributed to their survival here?
5. **GS5** There are three major types of rainforest *environments* found in Australia. What makes Queensland's rainforests unique? Why is this possible?
6. **GS5** Refer to **FIGURE 3**. Why are there no rainforest *environments* on the western side of Australia?
7. **GS2** Why do the Kuku Yalanji people regard damage to the Daintree Rainforest as unacceptable?

6.4 Exercise 2: Apply your understanding

1. **GS6** Based on the history of Australia's rainforests and the protection now in place for the remaining forests, what do you think the future holds for this important resource?
2. **GS6** Which of the present uses of the rainforest do you think is the most *sustainable* for the forest's future? Explain your answer.
3. **GS5** Refer to **FIGURE 8**. Why does the clearing and *change* in the Amazon appear to occur in straight lines?
4. **GS3** Looking at **FIGURES 6**, **7** and **8** for *interconnections*, what do you think could be contributing to the high levels of sediment in the Amazon River? Why?

5. **GS5** Refer to **FIGURE 9**.
   (a) Explain the role of the rainforest *environment* in relation to the climate.
   (b) Why are rainforests sometimes called 'the lungs of the Earth'?
6. **GS5** Look carefully at **FIGURE 5**.
   (a) List the countries of South America into which the Amazon rainforest extends.
   (b) Which country contains most of the Amazon rainforest?
   (c) Why do you think there are so few large cities in the rainforest?
   (d) Estimate the percentage of the rainforest that can be considered:
      i. under low or no threat
      ii. under threat
      iii. disturbed.
      Describe in your own words what each of these terms means.
7. **GS6** If development in the Amazon Basin continues as seen in **FIGURES 7** and **8**, what could be the consequences in terms of the processes shown in **FIGURE 9**?

Try these questions in learnON for instant, corrective feedback. Go to www.jacplus.com.au.

6.5 SkillBuilder: Drawing a précis map

What is a précis map?
A précis map is a simplified map — the cartographer has decided which details to leave in and which to leave out. It is different from a sketch map, which includes all the main features.

Select your learnON format to access:
- an overview of the skill and its application in Geography (Tell me)
- a video and a step-by-step process to explain the skill (Show me)
- an activity and interactivity for you to practise the skill (Let me do it)
- questions to consolidate your understanding of the skill.

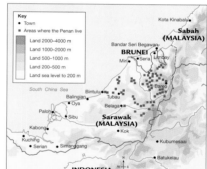

on Resources

Video eLesson SkillBuilder: Drawing a précis map (eles-1657)
Interactivity SkillBuilder: Drawing a précis map (int-3153)

6.6 Indigenous peoples and the rainforest

6.6.1 The Huli people of Papua New Guinea
It is difficult to accurately count all the people around the world who live in rainforests, but some estimates put the number as high as 150 million, including indigenous people. While these people are usually described as living a traditional **subsistence** way of life, this is generally combined with selling and buying items such as their labour, their land and assorted forest products.

Some 80 000 Huli people live in montane rainforest in the highlands of Papua New Guinea. The land on which they live has steep hillsides and dense rainforest. In the mountains the rivers cut deep gorges, and as they reach flat areas they form swampy, fertile basins.

The Huli people today use a farming system known as **shifting agriculture**. A patch of rainforest is cleared and crops of sweet potato, sugar cane, corn, taro and green vegetables are planted. It is the role of the women to tend these gardens, and their individual huts are built next to the gardens. The men live together in a communal house and generally look after themselves.

When the soil of the garden no longer produces good crops, a new patch of rainforest is cleared, leaving the old one to recover naturally. The garden crops are supplemented by food that the men obtain by hunting. Wild and domesticated pigs are a common source of meat.

While most Huli people still live on their lands, the influence of Western society is very obvious. Most Huli people wear some items of western-style clothing, and knives, cooking utensils and mirrors are common.

FIGURE 1 The Huli people of Papua New Guinea make wigs from their own hair, decorated with feathers from birds of paradise and colourful parrots.

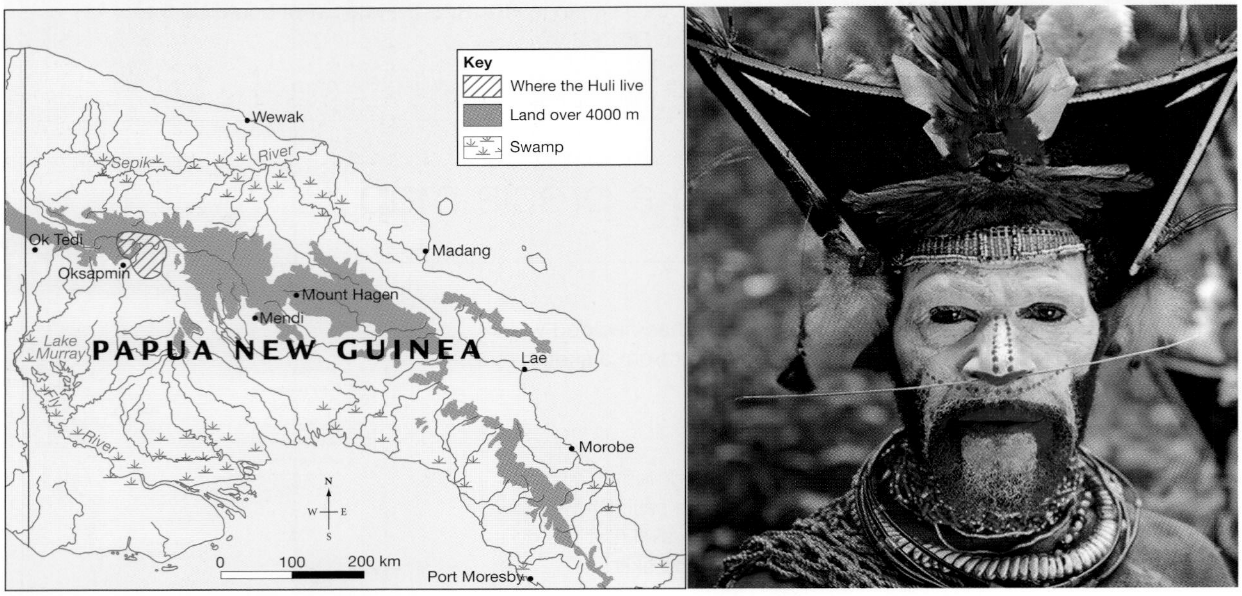

Source: MAPgraphics Pty Ltd, Brisbane

6.6.2 The Penan people of Borneo

The Penan people of Malaysian Borneo are a truly **nomadic** rainforest people. Although their forest home has been largely destroyed by logging, about a thousand Penan people remain deep in the forest, following their traditional way of life.

The main food sources for the Penan people are the sago palm and other fruiting trees, but they are also extremely skilled hunters. They use blowpipes and poison darts to kill wild pigs and gibbons. Their knowledge of the rainforest has been built up over thousands of years, and the forest provides for all their needs. They do not practise agriculture; instead, they follow the flowering cycle of the sago palm.

In order to survive in this environment, the Penan people have a strong culture of sharing. This applies not only to objects used in daily life, such as cooking utensils and blowpipes, but also to land. The idea of owning land does not exist in Penan culture.

The Penan people recently became well known to the outside world when they blockaded roads in the Malaysian rainforest to stop logging trucks and machinery moving into it.

DISCUSS
Imagine that you live in a society in which no-one recognises individual ownership of anything. In groups of three or four, discuss the advantages and disadvantages of living in this type of society.

[Creative and Critical Thinking Capability]

FIGURE 2 A Penan hunter of Sarawak or Brunei can shoot a dart 50 metres from their blowpipe. About 1.2 metres long and completely straight, a blowpipe has a very accurate hole through the middle.

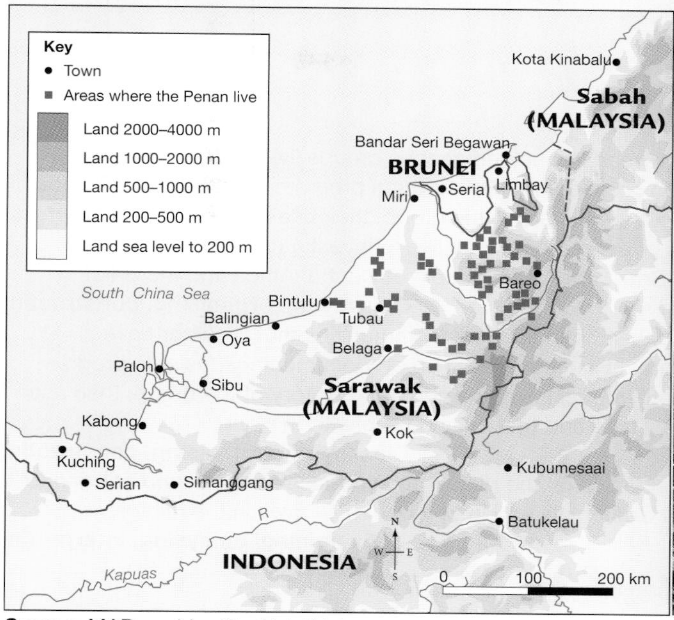

Key
- Town
- Areas where the Penan live

Land 2000–4000 m
Land 1000–2000 m
Land 500–1000 m
Land 200–500 m
Land sea level to 200 m

Kota Kinabalu

Sabah (MALAYSIA)

Bandar Seri Begawan
BRUNEI
Miri • Seria Limbay

South China Sea
Bintulu
Balingian Tubau
Belaga
Oya
Paloh
Sibu
Kabong
Kuching
Serian Simanggang

Bareo

Sarawak (MALAYSIA)

Kok

Kubumesaai

Batukelau

INDONESIA

Kapuas

0   100   200 km

Source: MAPgraphics Pty Ltd, Brisbane

6.6.3 The Korowai and Kombai people of Papua

The Korowai and Kombai peoples live in the Indonesian province Papua, in the south-western part of the island of New Guinea. Mosquitoes, floodwaters and community rivalries have forced these groups to build houses high up in the forest's canopy. They collect food (for example, sago) from the rainforest using tools such as stone axes.

FIGURE 3 The houses of the Korowai and Kombai peoples can be up to 40 metres high in the forest's canopy.

6.6 INQUIRY ACTIVITIES

1. Complete one of the following activities.
   (a) In groups of two or three, use the internet to research other indigenous groups living in rainforests around the world. Create a visual presentation or 'documentary' (using a program such as PowerPoint or Photo Story) to educate your classmates about the history of the group, their use of their **environment**, and threats or **changes** they face. Examples of peoples you may like to investigate include the Kuna people of Panama, the Yanomami people of Brazil, the Mbuti people of Central Africa and the Baka people of Cameroon. **Classifying, organising, constructing**
   (b) In groups of two or three, use the internet to research Indigenous communities that live in or have a connection to the rainforest in Australia. Create a visual presentation or 'documentary' (using a program such as PowerPoint or Photo Story) to educate your classmates about the history of the group, their use of their **environment**, threats or **changes** they face and their relationship with the land. **Classifying, organising, constructing**
2. Use the **Treehouse** weblink in your Resources tab (click on the picture then choose the *My world* activity) to see the community of Paso Caballos in northern Guatemala and learn about their lives in the rainforest. **Examining, analysing, interpreting**

6.6 EXERCISES

Geographical skills key: GS1 Remembering and understanding **GS2** Describing and explaining **GS3** Comparing and contrasting **GS4** Classifying, organising, constructing **GS5** Examining, analysing, interpreting **GS6** Evaluating, predicting, proposing

6.6 Exercise 1: Check your understanding

1. **GS3** Make a list of the similarities between the Huli people and the Penan people. Make another list of the differences between them. Use the photographs as well as the text in this section to help you.
2. **GS2** The Penan people use only rainforest resources to make their blowpipes. How do you think they do this? Use diagrams to illustrate your ideas.
3. **GS2** Why is a blowpipe better than a rifle in the rainforest?
4. **GS3** The Penan people are nomadic. How is this lifestyle different to the shifting agriculture practised by the Huli people?
5. **GS3** Describe how the life of men and women differ in the Huli society.

6.6 Exercise 2: Apply your understanding

1. **GS3** Explain what is meant by shifting agriculture and how it differs from the farming methods used in Australia.
2. **GS6** The Korowai and Kombai people build their homes high in the forest canopy. How do you think they do this? Use diagrams to illustrate your answer.
3. **GS6** What is the major threat to the traditional lifestyle of the indigenous people of the rainforest? Justify your answer.
4. **GS3** Compare the lifestyle of the Huli and the Penan people. Use a Venn diagram to explore the similarities and differences between them.
5. **GS6** Indigenous rainforest peoples still practise their traditional lifestyles rather than a western-style lifestyle. Suggest a reason for this.

Try these questions in learnON for instant, corrective feedback. Go to www.jacplus.com.au.

6.7 Disappearing rainforests

6.7.1 Factors causing rainforest deforestation

Rainforests have the potential to provide a wide variety of useful resources. The temptation to use these pristine areas is often too difficult for people to resist, especially if they live in poverty. As a result, all around the world, rainforests are being destroyed for economic gain. The main reasons for rainforests being cleared are described below.

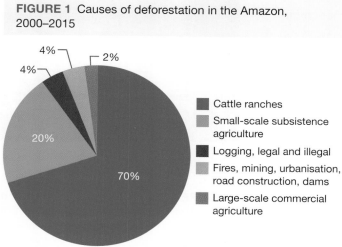

FIGURE 1 Causes of deforestation in the Amazon, 2000–2015

- 4%
- 2%
- 4%
- 20%
- 70%

■ Cattle ranches
■ Small-scale subsistence agriculture
■ Logging, legal and illegal
■ Fires, mining, urbanisation, road construction, dams
■ Large-scale commercial agriculture

Most recent data available at time of publishing

Commercial logging

There are two main types of logging: **clearfelling** and **selective logging**. When a forest is clearfelled, all trees are removed either by chainsaw or with heavy machinery such as bulldozers. In selective logging, only the best and most valuable trees are cut down. But in clearing forest to reach those trees, it is estimated that a hectare (10 000 square metres) of forest is destroyed for each log removed.

Farming

Rainforests grow in many developing countries. These countries struggle to provide the basic necessities of life for their people, and their populations are often rapidly increasing in size. In these countries, the land on which the forest grows is seen as more valuable than the forest itself.

FIGURE 2 It is thought that up to 80 per cent of logging in Brazil and Indonesia could be illegal.

Highways create access to these areas, opening up parts of the rainforest once almost impossible to reach. Soon after the roads are built, settlers (called homesteaders) arrive. Claiming a piece of the forest that borders the road, the homesteaders chop down a few trees as timber for fencing or a house, and then set fire to the rest.

Once the initial 'land rush' is over and all the land beside the roads has been claimed, tracks and roads leading from the highways will push deeper and deeper into the forest. Soon an area of 50 kilometres either side of the highway will have been destroyed and replaced by small farms or large-scale commercial farms that raise beef or crops for export to the richer countries of the world.

Mining

Many rainforests are growing on land that also contains large energy and mineral deposits such as oil, gold, silver, bauxite, iron ore, copper and zinc. Mineral companies build roads to the deposits and set up large-scale mining and processing plants. These plants require large amounts of electricity, and this is often supplied by burning trees to create charcoal or by constructing **hydroelectric dams**.

Deep in the Brazilian rainforest, a 2000-square-kilometre dam has been constructed to provide electricity for aluminium smelters. The dam flooded the entire tribal lands of two native peoples, and is so large that it has altered the climate in the area, making it drier.

Another problem created by mining is the pollution of nearby rivers and streams from chemicals used in the processing plants. Rivers downstream from a vast goldmine in Papua New Guinea have been found to contain four times the safe limit of cyanide in the water. Cyanide is used to extract gold from rock.

FIGURE 3 Blocks of rainforest in Peru are burned to clear the area for agricultural use — here, maize seedlings have been planted.

FIGURE 4 The Ok Tedi gold and copper mine in the Papua New Guinea rainforest. The damage that mining has caused to the surrounding environment can be clearly seen.

DISCUSS
As a class, discuss the potential long-term problems that could result from the continued commercial use of rainforest *environments* around the world. Develop a list of the top five potential problems. **[Ethical Capability]**

6.7 INQUIRY ACTIVITY
Using the internet, research any economic activities that are supported by Australian rainforests.
Examining, analysing, interpreting

6.8 Social and environmental impacts of deforestation

6.8.1 Impacts of rainforest deforestation

Deforestation of rainforests around the world is the major cause of problems in this ecosystem. The loss of unique **habitats** is the primary reason species are becoming endangered. Clearing creates smaller islands of vegetation, making it more difficult for animals to communicate and breed. People are also affected by the removal of the rainforest. While indigenous peoples may feel the effects first, others also experience negative consequences.

- About one hectare of rainforest is destroyed every second: this is about twice the size of a soccer pitch.
- Scientists estimate that 137 plants and animals are made extinct daily: that's 50 000 each year. Some haven't even been discovered yet!
- It is believed that in the year 1500 up to nine million indigenous peoples lived in the Amazon rainforest. The number is now lower than 200 000.
- The world loses about two per cent of its rainforest each year, but rates differ between countries.

6.8.2 Impacts on plants and animals

Islands in the forest

Many forests are cleared using fire. These fires will release millions of tonnes of carbon dioxide into the air, increasing the threat of global warming. At the same time, destroying the trees robs the planet of the natural system that helps regulate the amount of carbon dioxide in the air.

In many areas where forests are cleared, it has become a practice to leave behind 'islands' of rainforest. This is meant to assist in the natural regeneration of the forest and also to leave sufficient areas of the natural habitats of plants and animals that live in the rainforest. But is this working?

The islands that are left are often not big enough to ensure the survival of the large numbers of species that live there. For example, the endangered Queen Alexandra's Birdwing (the world's largest butterfly) is facing extinction. Confined to coastal rainforests near Popondetta Province in northern Papua New Guinea, its survival depends on the presence of old growth forests. Although the Popondetta covers approximately 100 square kilometres, butterfly populations are now only found in five isolated pockets of up to two square kilometres. These remaining refuges are threatened by surrounding palm oil plantations.

And there are other problems. When the forest is cleared, the exposed earth can quickly erode as the tree roots no longer hold the soil together, making the regrowth of vegetation slow. On steep slopes this can increase the risk of landslides, and sediments can flow into rivers.

During drought, the bare ground can become hot and barren. With the removal of the forest cover there is little moisture stored in the ground and a much lower rate of **evapotranspiration**. This in turn affects the water cycle, reducing the amount of rain that falls on the remaining islands of rainforest, and they quickly dry out.

6.8.3 CASE STUDY: Deforestation in Indonesia and the orangutan

Nearly 10 per cent of the world's rainforests and 40 per cent of all Asian rainforests are found in Indonesia. Less than half of Indonesia's original rainforest area remains. Much of this is in Kalimantan, on the island of Borneo. Forests have been cleared for timber, for plantation crops such as palm oil trees, and to make way for Indonesia's growing population, which is now more than 200 million. Fires lit to clear land in 1982 and 1997 resulted in wildfires that severely damaged large areas of rainforest in Kalimantan. Orangutans, Sumatran tigers and Javan hawk-eagles may disappear from Indonesia as their natural habitats disappear.

Orangutans are the largest tree-living mammals and the only great ape that lives in Asia. They survive only on the islands of Borneo and Sumatra. Current estimates are that orangutans have lost 80 per cent of their habitat in the last 20 years. In 1997–98, wildfires burned through nearly two million hectares of land in Indonesia, killing up to 8000 orangutans.

FIGURE 1 Leftover pockets of rainforest are at risk from reduced rainfall and cannot survive drought conditions.

(a) Rainforest trees are cleared, with 'islands' left for regeneration.

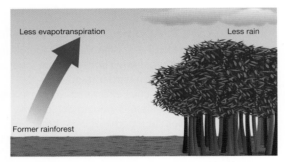

(b) There is less evapotranspiration and less rain on forest 'islands'.

FIGURE 2 The wingspan of the Queen Alexandra's Birdwing can reach 30 centimetres.

FIGURE 3 Mother and baby orangutan

It is estimated that orangutan numbers have declined by more than 50 per cent in the last 60 years. Hunting and killing, particularly on the island of Borneo, has resulted in the loss of 150 000 individuals in the last 16 years. The current orangutan population is estimated to be between 70 000 and 100 000. Conservationists predict that without a renewed effort to protect the species, by 2055 their numbers will decline by another 45 000.

FIGURE 4 Orangutan distribution in Borneo, 1930–2015

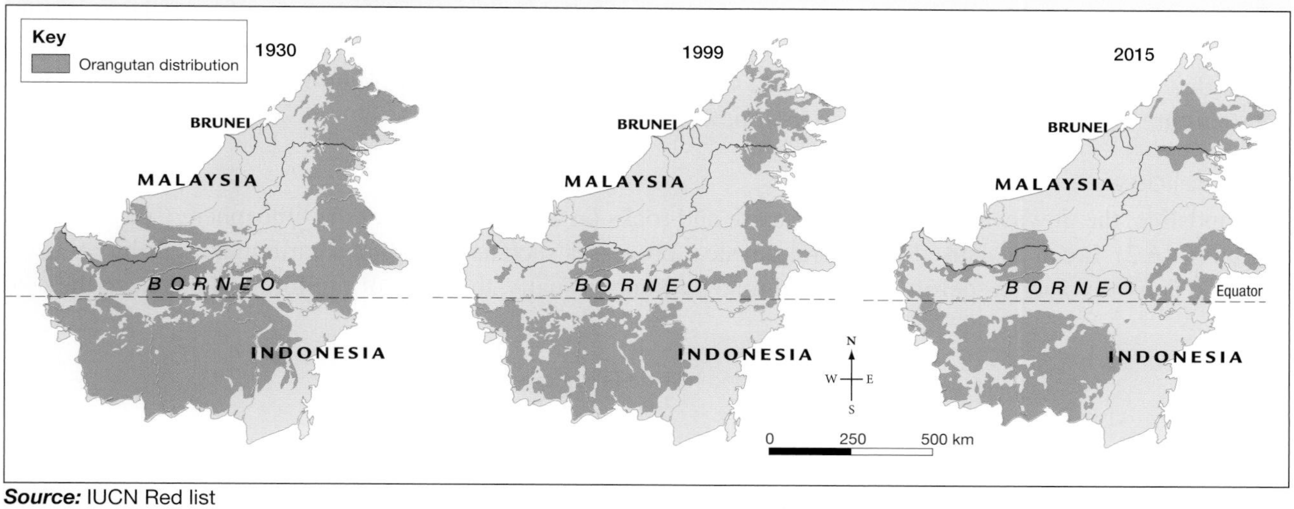

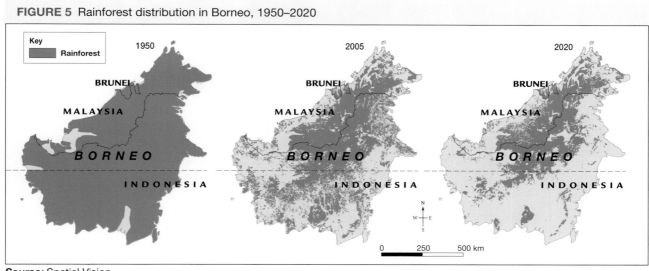

Source: IUCN Red list
Most recent data available at time of publishing

FIGURE 5 Rainforest distribution in Borneo, 1950–2020

Source: Spatial Vision

Explore more with my**World**Atlas

Deepen your understanding of this topic with related case studies and questions.
• Investigate additional topics > Endangered and introduced species > **Orangutans**

6.8.4 Impacts on people

Indigenous peoples

As forests are cleared and new occupiers move into the region, the indigenous peoples of the area are often displaced and their cultures may disappear. The homesteaders bring new diseases to which indigenous peoples have no natural immunity. One group, the Nambiquara of Brazil, lost half its population to illness when a road was placed through their tribal land. Indigenous peoples aren't often given a choice about 'progress' coming to their section of the rainforest. As a result, tension can be created between these indigenous communities and the government. In 1999, the Bakun Dam Project began in Malaysia, resulting in the eviction of approximately 10 000 indigenous people from their ancestral homeland. While they were resettled as compensation, the land provided was too small to support their traditional forms of hunting and agriculture and many failed to adapt to their new lifestyles.

Landslides

A landslide, the downward movement of earth and rocks on a slope, occurs in the lithosphere (see subtopic 6.2). It can be caused by natural physical processes such as rainfall and earthquakes, or by human activities such as deforestation and road building. Usually, the roots of rainforest plants keep the

FIGURE 6 A forested hillside (a) before and (b) after deforestation

(a)

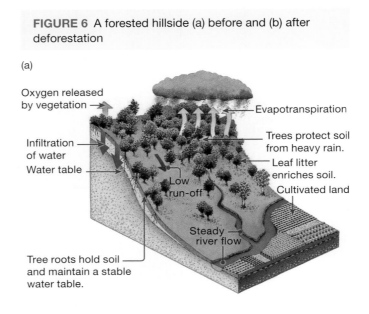

Oxygen released by vegetation
Evapotranspiration
Infiltration of water
Water table
Trees protect soil from heavy rain.
Leaf litter enriches soil.
Cultivated land
Low run-off
Steady river flow
Tree roots hold soil and maintain a stable water table.

(b)

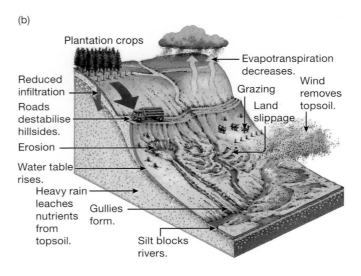

Plantation crops
Reduced infiltration
Roads destabilise hillsides.
Erosion
Water table rises.
Heavy rain leaches nutrients from topsoil.
Gullies form.
Silt blocks rivers.
Evapotranspiration decreases.
Grazing
Land slippage
Wind removes topsoil.

soil together and add stability to mountainous areas. This is especially important during times of heavier rainfall. However, sometimes the ground becomes so waterlogged that the roots can't keep the soil in place and it slips downhill, creating a landslide. The risk of this increases if deforestation has taken place on the hillside, as there are no tree roots to provide added stability.

Therefore, when these hills are cleared and settled by communities, the danger of property damage, and even death, increases. November 2011 saw 35 people killed in a landslide in the Colombian city of Manizales. Fourteen houses were destroyed, displacing up to 159 people. This mountainous, coffee-growing region used to be rainforest before it was cleared and settled.

The Philippines is at a high risk of landslides due to deforestation. Large tracks of forest have been removed by illegal logging, fires to clear the land for agriculture and mining operations. In 2018 Typhoon Ompong triggered massive landslides that buried the living quarters of miners. More than 100 people died despite attempts to clear the area before the Typhoon struck.

FIGURE 7 Landslides in intact forest in the Mata Atlantica rainforest, Brazil

Disease

The arrival of new tropical diseases is a less obvious result of deforestation. As animal **hosts** disappear and new human settlers move into previously inaccessible areas, 'new' disease-causing microorganisms are transferred into the human population. The frequency of mosquito-borne diseases such as malaria has increased due to the creation of more water puddles, for example in ditches and tyre treads, that are an excellent breeding ground for the mosquito. It is estimated that malaria is responsible for the deaths of 20 per cent of the Yanomami people in Brazil and Venezuela. Today, more than 99 per cent of malaria cases in Brazil occur in the Amazon Basin region, even though the mosquitoes that carry the disease are found across 80 per cent of the country. In 2018 the survival of Yanomami communities was further threatened by an outbreak of measles, thought to have been brought in by miners. With no natural immunity and a lack of medical care, whole populations can be wiped out.

FIGURE 8 Landslide in Manizales, Colombia, in November 2011

FIGURE 9 Deforestation and the subsequent erosion are clearly evident in Sumatra, Indonesia.

The outbreak of such diseases doesn't affect only the local area but the impact can also spread into other countries via people who visit these areas, unknowingly contract an illness and then travel home, spreading the disease along the way.

6.8 INQUIRY ACTIVITIES

1. Research and create a list of 10 other animal species threatened by deforestation around the world. Choose one of these animals and report back to the class on its current location, the remaining population level and the main causes of deforestation. Present your report as a poster, PowerPoint presentation, movie (documentary), poem, song or drama performance. **Classifying, organising, constructing**

2. Using the internet, investigate two different management strategies, policies or laws that have been implemented around the world to try to conserve the rainforest *environment*. Note the positive and negative aspects of these strategies. Comment on their ability to support the *sustainable* use of rainforests. Discuss your results as a class. Create a summary on the board to evaluate all the options that are shared. **Examining, analysing, interpreting**

3. Produce an A4-sized poster designed to publicise the rate and consequences of rainforest destruction. Your poster must include a colourful diagram and a short slogan based on the facts and figures presented in this subtopic. **Classifying, organising, constructing**

6.8 EXERCISES

Geographical skills key: GS1 Remembering and understanding **GS2** Describing and explaining **GS3** Comparing and contrasting **GS4** Classifying, organising, constructing **GS5** Examining, analysing, interpreting **GS6** Evaluating, predicting, proposing

6.8 Exercise 1: Check your understanding

1. **GS1** Name some species threatened by deforestation in Indonesia.
2. **GS1** List the main threats to orangutans.
3. **GS2** What is the *interconnection* between deforestation and the impact of disease on indigenous peoples?
4. **GS2** How does deforestation affect the lithosphere, atmosphere and biosphere? (Refer to subtopic 6.2 to refresh your memory.)
5. **GS2** Why does having separate small islands of vegetation make it more difficult for animals to communicate and breed?

6.8 Exercise 2: Apply your understanding

1. **GS3** Refer to **FIGURES 4** and **5**. Describe the *interconnection* between the two sets of data.
2. **GS2** Refer to **FIGURE 6**. Write a paragraph that explains how deforestation results in the consequences and *changes* illustrated in the diagram.
3. **GS6** Study **FIGURE 9**. Indonesia recently granted a licence to a pulp paper producer to clear 50 000 hectares of forest near an orangutan sanctuary in Sumatra. What impact do you consider this might have on the orangutan population?
4. **GS6** What could be some of the consequences if the rainforest *environment* continues disappearing at its current rate?
5. **GS2** Why is it important to save species from extinction?

Try these questions in learnON for instant, corrective feedback. Go to www.jacplus.com.au.

6.9 Saving and preserving rainforests

6.9.1 Options for conserving rainforests

As people begin to realise the importance of rainforests, many have started to work towards preserving these valuable 'green dinosaurs'. Some methods of conservation are relevant only to governments and large companies, but some are relevant to you and the choices you make.

Rescue package 1: protect the remaining rainforests

While only six per cent of the world's rainforests are in a national park or reserve, there are many large areas of rainforest under protection. The number and size of these national parks are slowly increasing. The Korup National Park in Cameroon holds 126 000 hectares of Africa's richest untouched rainforest; the Khao Yai National Park in Thailand has 200 000 hectares, where the habitats of tigers, elephants and gibbons are protected; Costa Rica's rainforests are the most protected of all, with national parks and reserves covering almost one-third of that country.

Rescue package 2: use the forest without destroying it

This is called **sustainable development**. It means that resources are taken from the rainforests but the forest remains largely intact. It has been estimated that a forest used this way is worth $12 000 a hectare, while it is worth only $300 a hectare if it is cleared for farming.

Timber users can now purchase timber from forests that are properly managed. A company in Mexico — the Forest Stewardship Council (FSC) — assesses forests around the world. If the forests comply with regulations, the timber is given the FSC stamp. People who purchase this timber know that the forest it came from is being responsibly managed.

FIGURE 1 The drill, one of Africa's most endangered primates, has a safe haven in the Korup National Park in Cameroon.

TABLE 1 Countries with FSC-certified forests totalling more than one million hectares, 2019

Country	Area of certified forest (hectares)
Australia	1 244 096
Belarus	8 957 566
Bosnia and Herzegovina	1 768 071
Brazil	7 085 315
Bulgaria	1 461 593
Canada	50 654 172
Chile	2 331 850
Congo, The Republic of	2 410 693
Croatia	2 048 581
Estonia	1 523 958
Finland	1 623 311
Gabon	1 741 228
Germany	1 357 027

(continued)

TABLE 1 Countries with FSC-certified forests totalling more than one million hectares, 2019 *(continued)*

Country	Area of certified forest (hectares)
Indonesia	2 626 297
Latvia	1 105 787
Lithuania	1 170 683
Mexico	1 338 522
New Zealand	1 248 195
Poland	6 955 564
Romania	2 838 745
Russia	46 764 362
South Africa	1 438 881
Sweden	13 370 511
Turkey	3 121 401
Ukraine	4 296 157
United Kingdom	1 637 196
United States	13 933 516

Rescue package 3: use alternative timber

One further step is to not use rainforest timber at all. Many rainforest trees are now grown in plantations, and alternatives such as using steel beams in houses and recycled paper in cardboard help take the strain off the rainforests.

One alternative that has been developed is the processing of old coconut palms to create hardwood. The company that is developing this resource, Tangaloa, claims that there are enough non-productive coconut palms to produce timber equivalent to one million rainforest trees. If this concept proves popular, plantations of coconut palms could be grown specifically for this purpose.

Rescue package 4: act now!

While most of us do not have rainforests growing in our backyards, the choices we make each day can and do make a difference to the way resources are used around the world. There are many organisations that aim to conserve the world's remaining rainforests. Some of their suggestions are:

- use less wood and paper
- write to businesses that destroy the rainforest
- educate yourself about the importance of rainforests
- look for alternatives to rainforest products
- be an **ecotourist** — visit rainforests where your tourist dollars go towards education and conservation.

FIGURE 2 Coconut plantation — could these palms help save the rainforests?

 Resources

Interactivity Protecting or plundering rainforests (int-3114)

6.9 INQUIRY ACTIVITIES

1. On a countries outline map of the world, shade in those countries with FSC-certified forests of over one million hectares. Use lighter shades of one colour for countries with smaller areas of certified forest (such as 1 000 000–2 499 999 and 2 500 000–4 999 999 hectares), and darker shades of the same colour for countries with larger areas (5 000 000–7 499 999; 7 500 000–9 999 999; >10 000 000 hectares). This type of map is called a choropleth map. **Classifying, organising, constructing**

2. Design your own website encouraging people to donate money to save the rainforest *environment*. **Evaluating, predicting, proposing**

3. Other methods to help conserve the world's rainforests include:
   • breeding endangered rainforest animals in captivity, and then releasing them
   • providing websites where sponsors can give money to buy some rainforest and put it into a reserve
   • employing indigenous people to pick nuts and berries or even to breed butterflies for collectors.

   Use the internet to find an example of each of these methods and list any others that you find while completing this research. Document your findings. **Examining, analysing, interpreting**

6.9 EXERCISES

Geographical skills key: GS1 Remembering and understanding **GS2** Describing and explaining **GS3** Comparing and contrasting **GS4** Classifying, organising, constructing **GS5** Examining, analysing, interpreting **GS6** Evaluating, predicting, proposing

6.9 Exercise 1: Check your understanding

1. **GS1** What percentage of the world's rainforests are in national parks or reserves?
2. **GS1** Which country has the most protected rainforests?
3. **GS1** How are rainforest *environments* in Costa Rica protected?
4. **GS2** Explain in your own words what the FSC does to help protect the rainforest *environment*.
5. **GS2** Explain what you understand by the term 'green dinosaur'.

6.9 Exercise 2: Apply your understanding

1. **GS6** List two advantages and two disadvantages of each rescue package listed in this subtopic. Which of the four packages do you think offers the most hope for rainforest conservation and *sustainability*? Explain why.
2. **GS2** Why is it good to have a variety of action options?
3. **GS3** Use an atlas to help you classify countries with FSC-certified rainforests over one million hectares by continent. Which continent has the most and which has the least FSC-certified forest?
4. **GS2** Write a letter to the editor, explaining the alternative timber products that are currently available.
5. **GS5** Explain why 'sustainable development' use of rainforest resources is more profitable than clearing the land for farming.

Try these questions in learnON for instant, corrective feedback. Go to www.jacplus.com.au.

6.10 Thinking Big research project: Rainforest display

SCENARIO

You have been commissioned by the Department of Natural Resources and Environment to complete an in-depth study on the importance of rainforests and present your information on their website, as part of an ongoing educational program.

Select your learnON format to access:

- the full project scenario
- details of the project task
- resources to guide your project work
- an assessment rubric.

 Resources

 projectsPLUS Thinking Big research project: Rainforest display (pro-0172)

6.11 Review

6.11.1 Key knowledge summary

Use this dot point summary to review the content covered in this topic.

6.11.2 Reflection

Reflect on your learning using the activities and resources provided.

 Resources

 eWorkbook Reflection (doc-31352)

Crossword (doc-31353)

 Interactivity Rainforest landscapes crossword (int-7599)

KEY TERMS

catchment area of land that drains into a river

clearfelling a forestry practice in which most or all trees and forested areas are cut down

compost a mixture of various types of decaying organic matter such as dung and dead leaves

drainage basin an area of land that feeds a river with water; or the whole area of land drained by a river and its tributaries

ecosystem an interconnected community of plants, animals and other organisms that depend on each other and on the non-living things in their environment

ecotourist a tourist who travels to threatened ecosystems in order to help preserve them

evapotranspiration the process by which water is transferred to the atmosphere from surfaces such as the soil and plants

gorge narrow valley with steep rocky walls

habitat the total environment where a particular plant or animal lives, including shelter, access to food and water, and all of the right conditions for breeding

host an organism that supports another organism

hydroelectric dam a dam that harnesses the energy of falling or flowing water to generate electricity

ice ages historical periods during which the Earth is colder, glaciers and ice sheets expand and sea levels fall

leaching a process that occurs in areas of high rainfall, where water runs through the soil, dissolving minerals and carrying them into the subsoil. The process can be compared to a coffee pot in which the water drips through the coffee grounds.

microclimate specific atmospheric conditions within a small area

nomadic describes a group that moves from place to place depending on the food supply, or pastures for animals

selective logging a forestry practice in which only selected trees are cut down

shifting agriculture process of moving gardens or crops every couple of years because the soils are too poor to support repeated sowing

species a biological group of individuals having the same common characteristics and being able to breed with each other

subsistence producing only enough crops and raising only enough animals to feed yourself and your family or community

sustainable development economic development that causes a minimum of environmental damage, thereby protecting the interest of future generations

temperate describes the relatively mild climate experienced in the zones between the tropics and the polar circles

UNIT 2
CHANGING NATIONS

Have you ever stopped to consider why you live where you do? What prompted your family to live there? There are so many different types of places where you *could* live: rural or urban, coastal or inland, small or large, bustling or quiet. Different people find different places suitable (or more 'liveable') for them than other places. Some people have no choice. The question is: how can we make places more liveable?

GEOGRAPHICAL INQUIRY: INVESTIGATING AN ASIAN MEGACITY

on line only

Your task

Your team has been put in charge of creating a website designed to inform the residents of an Asian megacity about its characteristics. Each city will be different depending on its location, wealth or poverty, size and climate. A key feature of your website will be to cover any urban solutions and innovations that are currently being implemented in your megacity.

Select your learnON format to access:
- an overview of the project task
- details of the inquiry process
- resources to guide your inquiry
- an assessment rubric.

on Resources

💡 **ProjectsPLUS** Geographical inquiry: Investigating an Asian megacity (pro-0146)

7 Urbanisation and people on the move

7.1 Overview

Why do millions of people choose to live so close to other people in busy urban areas?

7.1.1 Introduction

There are many advantages to living in large cities — for example, the economic benefit brought about by sharing the costs of providing fresh water, electricity or other energy sources and public transport between many people. There may be social benefits, because the cities provide a wider choice of sporting, recreational and cultural events. However, there are also disadvantages of living in a large city environment. In this topic we will explore and compare urbanisation around the world.

 Resources

☑ **eWorkbook** Customisable worksheets for this topic

⊞ **Video eLesson** Our urban world (eles-1628)

LEARNING SEQUENCE

7.1 Overview

7.2 Urbanisation around the world

7.3 Australian urbanisation

7.4 **SkillBuilder:** Understanding thematic maps `online only`

7.5 Comparing urbanisation in the United States and Australia

7.6 Effects of international migration on Australia

7.7 **SkillBuilder:** Creating and reading pictographs `online only`

7.8 People on the move in Australia and China

7.9 **SkillBuilder:** Comparing population profiles

7.10 **Thinking Big research project:** Multicultural Australia photo essay `online only`

7.11 **Review** `online only`

To access a pre-test and starter questions and receive immediate, **corrective feedback** and **sample responses** to every question, select your learnON format at www.jacplus.com.au.

7.2 Urbanisation around the world

7.2.1 What is urbanisation?

As the world's population increases, **urban** areas continue to grow. In some regions, people are moving from rural to urban areas at very high rates.

Urbanisation is the growth and expansion of urban areas and involves the movement of people to towns and cities. The earliest cities emerged about 5000 years ago in Mesopotamia (part of present-day Iran, Iraq and Syria). Originally these cities depended on agriculture. In 1800, 98 per cent of the global population lived in rural areas and most were still dependent upon farming and livestock production — only 2 per cent of people lived in urban areas.

However, as cities grew and trade developed, urban areas became centres for merchants, traders, government officials and craftspeople. By 2008, the proportion of people living in urban areas had increased to 50.1 per cent, and in 2017 the figure had risen again to nearly 55 per cent. The rate of growth has varied in different regions (see **FIGURE 1**).

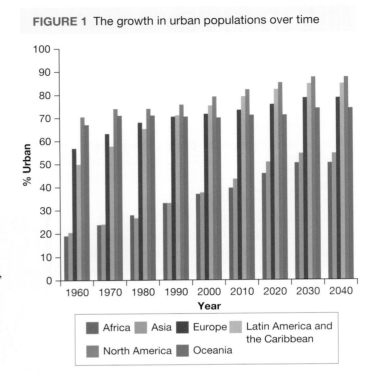

FIGURE 1 The growth in urban populations over time

Legend: Africa, Asia, Europe, Latin America and the Caribbean, North America, Oceania

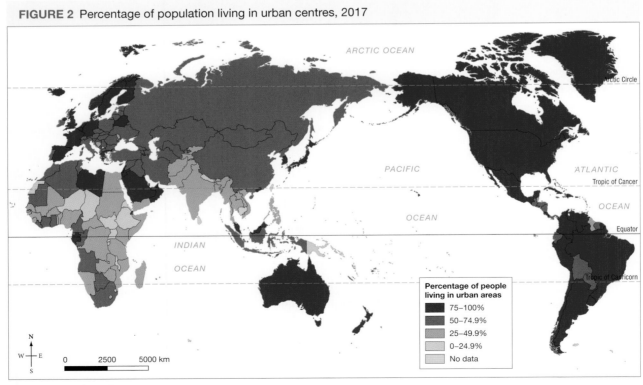

FIGURE 2 Percentage of population living in urban centres, 2017

Percentage of people living in urban areas
- 75–100%
- 50–74.9%
- 25–49.9%
- 0–24.9%
- No data

Source: World Bank Data

7.2.2 Uneven urbanisation

Urbanisation has not occurred evenly across the world. Some countries are predominantly rural, such as Cambodia and Papua New Guinea (populations 77 per cent and 87 per cent rural respectively), whereas others are almost completely urban, such as Belgium and Kuwait (98 per cent urban for both). In fact, some countries have 100 per cent urbanisation, including Bermuda, Cayman Islands, Hong Kong, Macau, Monaco, Vatican City and Singapore. South America is becoming one of the most urbanised regions in the world and currently has a population of around 385 million people. It is estimated that by 2050, 91.4 per cent of its population will be residing in urban areas.

Coastal urbanisation

People have lived on coastlines for thousands of years. Often at the mouth of rivers, coastal settlements became centres of trade and commerce and quickly grew into cities. Today, about half the world's population lives along or within 200 kilometres of a coastline (see **FIGURE 4**). According to the European Commission, 95 per cent of the world's population lives on only 10 per cent of the Earth's land area.

Countries that have over 80 per cent of their population living within 100 kilometres of a coastline include the United Kingdom, Senegal, Portugal, Belgium, the Netherlands, Sweden, Norway, Tunisia, Greece, Oman, the United Arab Emirates, Kuwait, Qatar, Sri Lanka, Japan, Singapore, Indonesia, Malaysia, the Philippines, Australia and New Zealand.

FIGURE 3 Urban housing in Kuwait

FIGURE 4 Cape Town in South Africa is a city located on the coast.

 **Resources**

Interactivity Urban Indonesia (int-3115)

Google Earth Cape Town

Indonesia

Explore more with my**World**Atlas

Deepen your understanding of this topic with related case studies and questions.
- Investigate additional topics > Urbanisation > **World urbanisation**

7.2 INQUIRY ACTIVITIES

1. Refer to a world population density map in your atlas or online. Compare this map with the two regions that have the highest rural population. What pattern do you see? **Comparing and contrasting**
2. Look at a physical map in an atlas to locate the countries with more than 80 per cent of their population located on the coast. Study the location of each country and create a table to record possible reasons for this pattern. **Classifying, organising, constructing**

7.2 EXERCISES

Geographical skills key: GS1 Remembering and understanding **GS2** Describing and explaining **GS3** Comparing and contrasting **GS4** Classifying, organising, constructing **GS5** Examining, analysing, interpreting **GS6** Evaluating, predicting, proposing

7.2 Exercise 1: Check your understanding

1. **GS1** Define *urbanisation* in your own words.
2. **GS2** How has urbanisation changed from 1960 to the present? How is this different around the world?
3. **GS2** What is expected to happen with urbanisation in the future?
4. **GS2** Explain how **FIGURE 1** shows that urbanisation has varied in different regions of the world. Which two regions have the greatest rural population?
5. **GS1** Look at **FIGURE 1**. Which region's urbanisation rate has consistently been the highest over time?

7.2 Exercise 2: Apply your understanding

1. **GS5** Look at **FIGURE 2**, which shows the population in urban areas. Identify and name the three countries with the highest and the three with the lowest percentage of people living in urban areas. Write a description of the general pattern shown in the map. Include patterns within different continents in your description.
2. **GS6** Rural areas are where most food is produced. What are two possible outcomes for food production if urbanisation continues?
3. **GS4** Draw a sketch of the photograph of Cape Town in **FIGURE 4**. Annotate the sketch, identifying the possible advantages and disadvantages to the natural *environment* when cities and towns are located on the coast.
4. Look at **FIGURE 2**. How does Australia's urbanisation rate compare with it's closest neighbours?
5. Look at **FIGURE 1**. Which two continents have the lowest urbanisation rates?

Try these questions in learnON for instant, corrective feedback. Go to www.jacplus.com.au.

7.3 Australian urbanisation

7.3.1 Where do most Australians live?

Australians live on the smallest continent and in the sixth largest country on Earth. With a population of almost 25 million in 2019 and an area of 7 690 000 square kilometres, our **population density** is 3.1 people per square kilometre. We may think of ourselves as an outback-loving, farming nation, but we mostly live near the coast.

Most Australians currently live within a narrow coastal strip that extends from Brisbane in the north to Adelaide in the south. While 71 per cent of Australians live in major cities, one in ten people live in small towns of less than 10 000 people. In 2016 there were just over 1000 towns with populations of fewer than 1000. About 85 per cent of people live within 50 kilometres of the coast. Australians love the beach, but is it just a coastal location that can explain this uneven **population distribution** pattern?

FIGURE 1 Distribution of Australia's towns by population size, 2016

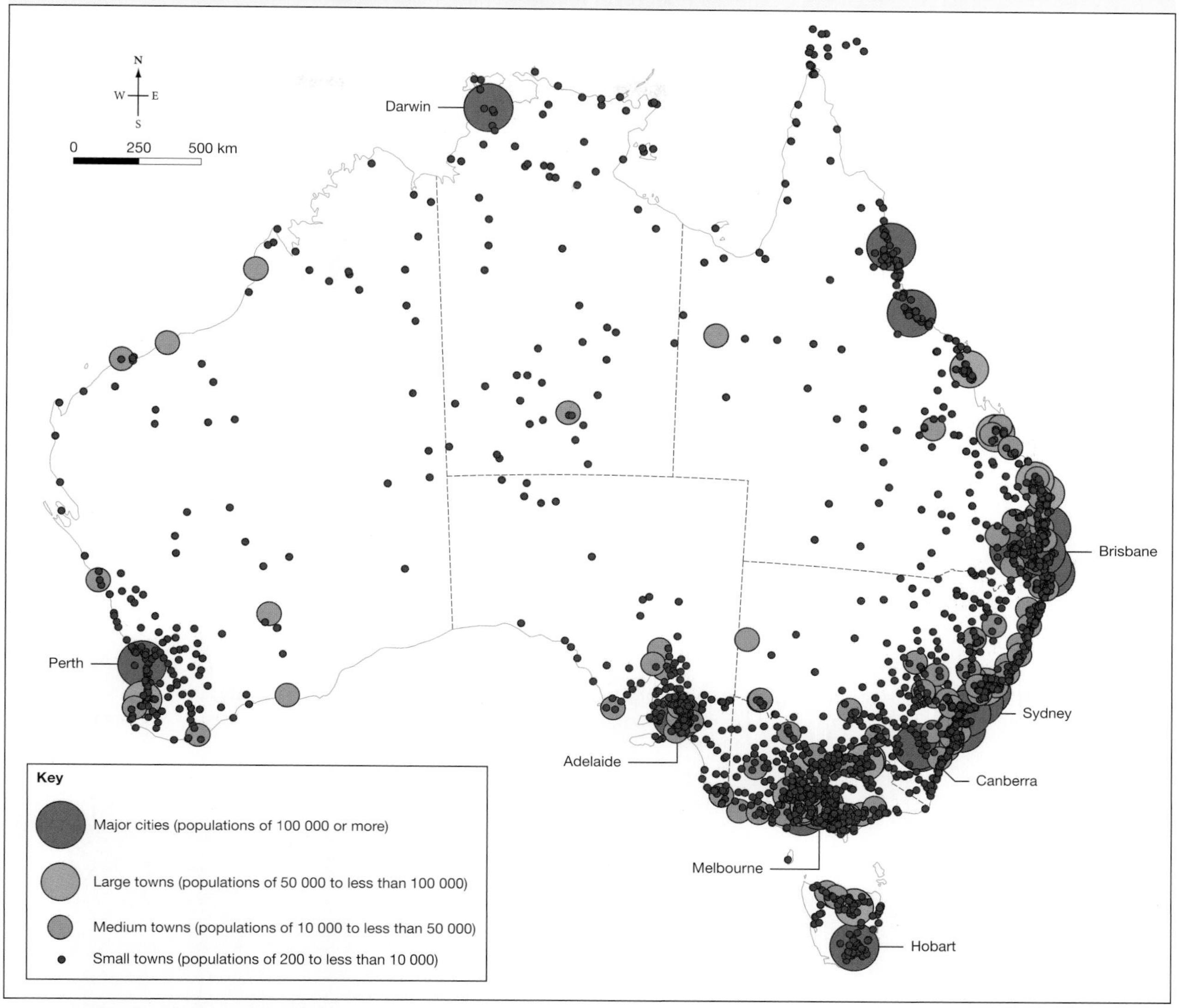

Key

- Major cities (populations of 100 000 or more)
- Large towns (populations of 50 000 to less than 100 000)
- Medium towns (populations of 10 000 to less than 50 000)
- Small towns (populations of 200 to less than 10 000)

Source: Australian Bureau of Statistics

─ Explore more with my**World**Atlas ──────────────

Deepen your understanding of this topic with related case studies and questions.
- Investigate additional topics > Population > **Population of Australia**

FIGURE 2 shows the distribution of rainfall within Australia. Comparing **FIGURES 1** and **2**, it is apparent that there is a strong interconnection between the availability of more than 800 millimetres of rainfall per year and population distribution in the east, south-east and south-west of Australia. It would be easy to say that Australians live in places where rainfall is higher, but if you look at these maps carefully there are major exceptions to this spatial pattern. What is the relationship between population distribution and total rainfall in the north of Australia? Is the population distribution high in the regions of high rainfall in Queensland and the Northern Territory?

Coastal locations and rainfall are not the only reasons Australians live where they do. The availability of mineral resources, irrigation schemes to enhance farm production, and remote and stunning tourist destinations are **geographical factors** that draw people to live in a particular place.

FIGURE 2 The distribution of annual rainfall in Australia

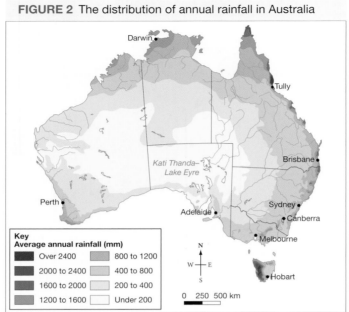

Source: Map drawn by MAPgraphics Pty Ltd, Brisbane

FIGURE 3 A remote town in northern Australia, which has a very low population density

7.3.2 Comparing population densities

FIGURE 1 shows Australia's population distribution in 2016. To better understand this data, we need to compare Australia's population density with that of other places in the world. This map shows that small areas around the major state capital cities have population densities of over 100 people per square kilometre of land. Look at **TABLE 1** and you can see that the average population density for Australia is well below the global average, and is easily the lowest of any of the permanently inhabited continents.

The population density of Australia is similar to that of Canada (3 people per square kilometre), but much lower than that of New Zealand (15 people per square kilometre), the United States (33 people per square kilometre) or China (145 people per square kilometre).

TABLE 1 The average population density for each continent

Continent	Average population density (people per km²)
Asia	100
Europe	55
Africa	36
North America	20
South America	32
Australia	3
Antarctica	0.00007

Consider the geographical factors that Australia might share with Canada but not New Zealand, the United States or China that could explain the significant difference between their population densities.

7.3.3 Where have Australians lived in the past?

Before European occupation

Prior to European arrival to Australia, where did **Indigenous** Australian peoples live?

Until 1788, Indigenous Australian peoples inhabited all parts of Australia (see **FIGURE 4**). The most densely populated areas, with 1–10 square kilometres of land per person, were the south-east, south-west and far north coastal zones, the north of Tasmania and along the major rivers of the Riverina region (south-western New South Wales).

The population density of Aboriginal and Torres Strait Islander peoples was highest in places close to coastal and river environments. These places had the best availability of food and other resources. In a location such as Port Jackson, New South Wales, food was abundant, meaning that the inhabitants needed to spend only about four hours each day hunting or gathering enough for their survival. In places where rainfall was unreliable, such as central Australia, the local peoples found it harder to survive. They often needed more than half a day to hunt and gather enough to satisfy their basic needs. When food resources ran low or with changing seasons, communities moved on to another part of their **country**. Being nomadic, they could manage their environment by not over-using the resources available at any one site.

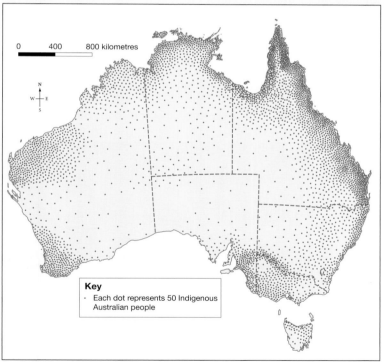

FIGURE 4 Where Aboriginal and Torres Strait Islander peoples lived in 1788

Key
· Each dot represents 50 Indigenous Australian people

Source: Map drawn by MAPgraphics Pty Ltd, Brisbane

7.3.4 Where do Aboriginal and Torres Strait Islander peoples live today?

It is believed that in 1788 there were between 350 000 and 700 000 Aboriginal and Torres Strait Islander peoples. Within 50 years this population had been greatly reduced by disease and British colonists. In 2016, there were 649 200 Aboriginal and Torres Strait Islander peoples, making up about 2.8 per cent of Australia's population.

The Australian environment has changed significantly since 1788. Much land has been cleared, shaped and blasted for cities, farms and mines. Other than the management of vegetation by fire, prior to European colonisation the landscape of Australia had not been greatly altered by its human inhabitants. By the twenty-first century, little of Australia's environment remained significantly unchanged by human occupation.

The patterns in **FIGURES 4** and **5**, showing the distribution of Aboriginal and Torres Strait Islander populations

FIGURE 5 Where Aboriginal and Torres Strait Islander peoples live today

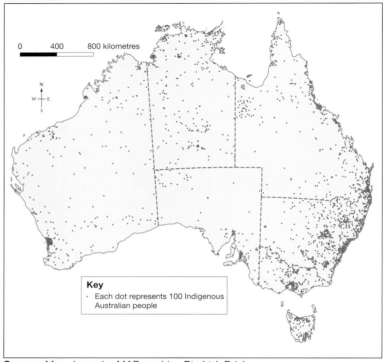

Key
· Each dot represents 100 Indigenous Australian people

Source: Map drawn by MAPgraphics Pty Ltd, Brisbane

in 1788 and today, are generally very similar. Since before 1788, most of Australia's peoples have tended to live in the same relatively small region of this country.

FIGURE 6 Many Aboriginal and Torres Strait Islander families enjoy living in remote parts of the country.

FIGURE 7 Regional distribution of Aboriginal and Torres Strait Islander peoples and the non-Aboriginal and Torres Strait Islander population of Australia

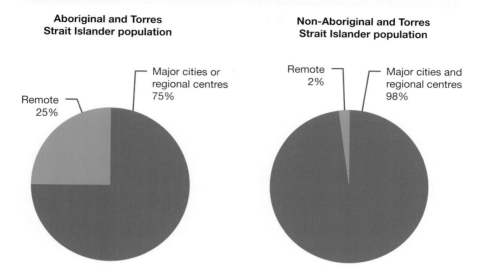

Aboriginal and Torres Strait Islander population

Remote 25%

Major cities or regional centres 75%

Non-Aboriginal and Torres Strait Islander population

Remote 2%

Major cities and regional centres 98%

FIGURE 8 The eight main climatic zones of Australia

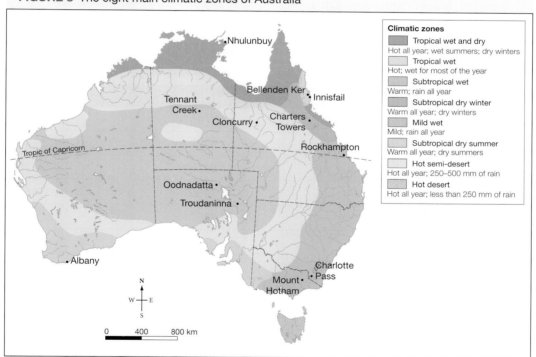

Climatic zones

Tropical wet and dry
Hot all year; wet summers; dry winters

Tropical wet
Hot; wet for most of the year

Subtropical wet
Warm; rain all year

Subtropical dry winter
Warm all year; dry winters

Mild wet
Mild; rain all year

Subtropical dry summer
Warm all year; dry summers

Hot semi-desert
Hot all year; 250–500 mm of rain

Hot desert
Hot all year; less than 250 mm of rain

Nhulunbuy
Bellenden Ker
Innisfail
Tennant Creek
Cloncurry
Charters Towers
Rockhampton
Tropic of Capricorn
Oodnadatta
Troudaninna
Albany
Charlotte Pass
Mount Hotham

N
W — E
S

0   400   800 km

Source: Map drawn by Spatial Vision

Explore more with my**World**Atlas

Deepen your understanding of this topic with related case studies and questions.
- Investigate additional topics > Population > **Indigenous Australians**

7.3.5 Is Australia an urbanised country?

With a population of nearly 25 million people in 2019 and a very large landmass, Australia has an average population density of only 3.1 people per square kilometre. Yet 85 per cent of people live within 50 kilometres of the coast, and most of these people — in 2018, 90 per cent of Australians — live in urban areas.

Australia is one of the most urbanised and coast-dwelling populations in the world and the level of urbanisation is increasing. From Federation (1901) until 1976, the number of Australians living in capital cities increased gradually from a little over one-third (36 per cent) to almost two-thirds (65 per cent). Since 1977, the population in capital cities has grown to 66 per cent. It is estimated that by 2053 this will have grown to 72 per cent (with an estimated 89 per cent in the four largest capital cities).

All of Australia's capital cities have grown over time, as have many regional urban areas such as the Gold Coast and Moreton Bay regions. This growth is expected to continue in the future (see **TABLE 2**).

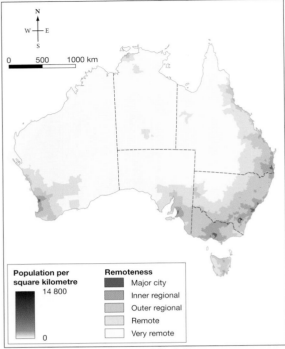

FIGURE 9 A map of Australia's population distribution shows that it is highly urbanised and coastal

Source: © Australian Bureau of Statistics

TABLE 2 Australian capital city 2017 populations and projected 2036 and 2066

City	2017 population	Projected 2036	Projected 2066
Sydney	5 132 355	7 379 976	11 240 860
Melbourne	4 843 781	7 520 830	12 235 490
Brisbane	2 413 457	3 596 431	5 782 256
Perth	2 039 041	2 798 994	4 330 509
Adelaide	1 334 167	1 605 335	2 068 550
Hobart	229 088	297 085	466 752
Darwin	148 884	195 082	295 458
Total	16 140 773	23 393 733	36 419 875

DISCUSS

Consider the issues and problems that increasing city populations will create. Discuss this as a class and construct a consequence chart to summarise all the ideas. What might be some solutions to these issues and problems? Add these to your chart. **[Critical and Creative Thinking Capability]**

7.3.6 What are the consequences of a highly urbanised Australia?

More land is needed when cities expand and this results in the greatest change — from agricultural to urban land. This has been called **urban sprawl**. Melbourne's growth has resulted in many new suburbs and extensive growth into and over food-growing areas, particularly in the west and south-east of the CBD (see **FIGURE 10**). Sydney, Perth and Brisbane have also spread into distant, previously agricultural areas.

FIGURE 10 Melbourne's urban growth over time

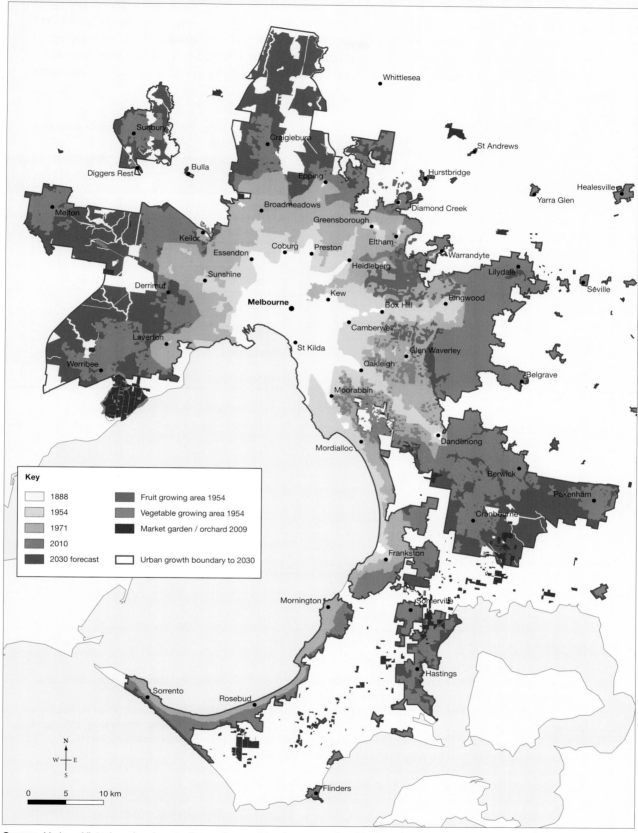

Key

	1888		Fruit growing area 1954
	1954		Vegetable growing area 1954
	1971		Market garden / orchard 2009
	2010		Urban growth boundary to 2030
	2030 forecast		

Source: Various Victorian planning studies and current land use mapping. Map produced by Spatial Vision 2019.

Historically, urban areas were settled where the land was flat, the water and soil were good and the climate was temperate — in other words, where good farmland is located. When cities spread, the sprawl takes over arable land (land able to be farmed for crops). Urban sprawl has long-term effects, as it is very difficult to bring the soil back to its former state once the predominant land use has been for buildings.

Many of Australia's cities have been called 'car cities' due to the reliance on cars and road networks for transport. These have an impact on distances and commuting times for people travelling to and from workplaces.

7.3.7 Ecological footprint

The amount of productive land needed on average by each person (in the world or in a country, city or suburb, for example) for food, water, transport, housing and waste management is known as an **ecological footprint**. It is measured in hectares per person per year. In 2016, the World Wide Fund for Nature (WWF) reported that the average global ecological footprint was 2.8 hectares per person. In 2014, Australia had an ecological footprint of 6.9 hectares per person. The United States had an ecological footprint of 8.4 hectare per person in 2014.

TABLE 3 Ecological footprints of Australian capital cities

City	Ecological footprint value (hectares/person/year)
Perth	7.66
Canberra	7.09
Darwin	7.06
Brisbane	6.87
Sydney	6.82
Adelaide	6.72
Melbourne	6.33
Hobart	5.50

on Resources

🔗 **Weblinks**   UAE ecological footprint

ABS: Indigenous health

Explore more with myWorldAtlas

Deepen your understanding of this topic with related case studies and questions.
• Investigate additional topics > Urbanisation > **Urbanisation in Australia**

7.3 INQUIRY ACTIVITIES

1. Use your atlas to identify and list:
   (a) geographical land forms or climatic features that are common to Australia and Canada. *Hint:* Look for large regions that have an extreme climate. Explain why.
   (b) reasons New Zealand, the United States or China may have a higher population density than Australia. Explain. **Examining, analysing, interpreting**
2. Refer to **FIGURES 1** and **2** to produce an overlay map that identifies the *interconnection* between the distribution of population and the distribution of rainfall within Australia.
   (a) Describe areas where there are strong similarities between these two features, i.e. high population distribution and high rainfall, or low population distribution and low rainfall.
   (b) Describe *places* that have a high population distribution but low rainfall or vice versa.
   Classifying, organising, constructing

3. Use various theme maps of Australia in your atlas to identify at least four possible *place* or *environmental* explanations for the pattern of distribution and density of Australia's population. Discuss your findings with the class.
   Examining, analysing, interpreting
4. Refer to **FIGURE 7**. Living so far away from major cities means that 25 per cent of Aboriginal and Torres Strait Islander communities have limited access to many of the services and opportunities that cities offer their residents. In a small group, brainstorm the lifestyle and service difficulties that may be associated with living so remotely.
   Evaluating, predicting, proposing
5. Collect some statistics that identify the health, wealth and educational inequalities that exist between Aboriginal and Torres Strait Islander peoples and non-Aboriginal and Torres Strait Islanders. For example, Aboriginal males have a life expectancy 17 years lower than that of non-Aboriginal males born in the same year. Use the **ABS: Indigenous health** weblink in the Resources tab to start your research. Write a paragraph or produce a series of graphs to comment on the inequalities you have discovered.
   Examining, analysing, interpreting
6. Conduct research to find which country in the world has the highest average population density. Find one country with a lower average population density than Australia.
   Comparing and contrasting
7. Use your atlas or online research to find an urban growth map for the capital city in your state or territory. Describe the *change* that has taken place over time. Using this map and a physical map of your state or territory, predict where future growth might occur. Justify your responses.
   Evaluating, predicting, proposing
8. (a) What is an ecological footprint?
   (b) Refer to **TABLE 3**. How does the ecological footprint data compare for Australian cities?
   (c) How do these figures compare with the average global ecological footprint?
   (d) Use internet sources (such as the **UAE ecological footprint** weblink in the Resources tab) to find out how the ecological footprint in the United Arab Emirates compares to that of Australian cities. What would happen if all cities had such a high footprint?
   (e) Create your own advertisement or animation using a video editing program to encourage people in your capital city to reduce their ecological footprint.
   Examining, analysing, interpreting

7.3 EXERCISES

Geographical skills key: GS1 Remembering and understanding **GS2** Describing and explaining **GS3** Comparing and contrasting **GS4** Classifying, organising, constructing **GS5** Examining, analysing, interpreting **GS6** Evaluating, predicting, proposing

7.3 Exercise 1: Check your understanding

1. **GS5** Use **FIGURE 1** and an atlas to describe where most people in Australia live.
2. **GS1** What is the difference between population density and population distribution?
3. **GS2** What geographical factors other than rainfall may lead to the uneven distribution of population in Australia?
4. **GS1** How many Aboriginal and Torres Strait Islander peoples:
   (a) lived in Australia in 1788
   (b) live in Australia today?
5. **GS1** What percentage of Australians live in urban areas? Of these, what percentage live in urban areas close to the coast?
6. **GS2** List the disadvantages of urban sprawl.
7. **GS2** Describe the growth of Melbourne over time. What impact has this growth had on food production areas?

7.3 Exercise 2: Apply your understanding

1. **GS4** Use the statistics in **TABLE 1** to produce a world map that illustrates the contrasts between the average population densities for each continent. *Hint:* A pictograph may best highlight the differences.
2. **GS6** Write a paragraph to explain the possible *change* in the distribution of Australia's predominantly urban population over the next 50 years if one of the following situations occurs.
   (a) The coastal urban areas become adversely affected by loss of land due to rising sea levels.
   (b) A 20-year-long drought occurs in south-eastern Australia.
3. **GS6** Use information from **FIGURE 2** to explain why, in the future, there may be significant movement of people from the southern states of Australia to *places* in the tropical north. Your answer must refer to specific information from the map.

▶

4. Study **FIGURES 1** and **8**.
   (a) **GS5** Identify the climatic zones in **FIGURE 8** that best match the population density areas in **FIGURE 1**.
   (b) **GS2** For each of the states shown in **FIGURE 8**, write a sentence to describe the climate for the region. For example, 'This region has a mostly mild to subtropical climate with rainfall all year round'.
5. **GS5** Refer to **FIGURE 9** and describe the population distribution of Australia.
6. **GS4** Refer to **TABLE 2**. Draw a bar graph to show the predicted **change** in the populations of Australia's capital cities. What does your graph reveal?

Try these questions in learnON for instant, corrective feedback. Go to www.jacplus.com.au.

7.4 SkillBuilder: Understanding thematic maps

online only

What is a thematic map?

A thematic map is a map drawn to show one aspect; that is, one theme. For example, a map may show the location of vegetation types, hazards or weather. Parts of the theme are given different colours or, if only one idea is conveyed, symbols may show location.

Select your learnON format to access:

- an overview of the skill and its application in Geography (Tell me)
- a video and a step-by-step process to explain the skill (Show me)
- an activity and interactivity for you to practise the skill (Let me do it)
- questions to consolidate your understanding of the skill.

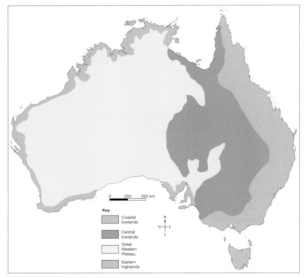

Key
Coastal lowlands
Central lowlands
Great Western Plateau
Eastern highlands

on Resources

Video eLesson SkillBuilder: Understanding thematic maps (eles-1658)

Interactivity SkillBuilder: Understanding thematic maps (int-3154)

7.5 Comparing urbanisation in the United States and Australia

7.5.1 Urbanisation in the United States and Australia

Both the United States and Australia are very large countries that are highly urbanised. In fact, both are among the world's most urbanised nations.

The United States and Australia have some similarities and some differences in terms of how urbanised they are, as revealed in **TABLE 1** and **FIGURE 1**.

TABLE 1 A comparison of urbanisation in the United States and Australia

	United States	Australia
Population	326 700 000 in 2018	24 530 000 in 2018
Population distribution	Over 81% live in urban areas, and 19.5% in rural areas.	Over 89% live in urban areas, less than 11% in rural areas.
People living in large cities	The United States has 10 cities that have a population of more than 1 million people.	Australia has 5 cities that have a population of more than 1 million people.
	Approximately 1 of every 10 people in the United States live in either the New York or Los Angeles metropolitan areas.	Approximately 4 of every 10 people in Australia live in either Melbourne or Sydney.

FIGURE 1 Population of the top 10 urban settlements in (a) the United States (2018) and (b) Australia (2019)

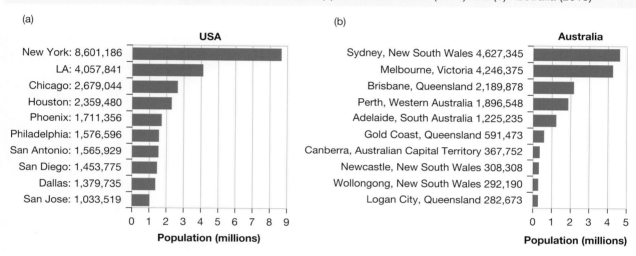

7.5.2 Causes of urbanisation

The causes of urbanisation are similar for both Australia and the United States. In each case, since the country was founded:

- fewer people were needed to work in rural areas as technology reduced the demand for labour on farms
- more jobs and opportunities were available in factories, which were located in urban areas
- the development of railways allowed goods produced in one city to be transported to rural and urban areas
- cities could grow and develop thanks to new technologies (steel-framed skyscrapers) and utilities (for example, electricity and water supply).

7.5.3 Consequences of urbanisation

Conurbations

Sometimes there are so many cities in a particular region that they seem to merge almost into one city as they expand. A conurbation is made up of cities that have grown and merged to form one continuous urban area. Both the United States and, to a lesser extent, Australia have conurbations.

United States

Eleven conurbations have been identified in the United States (see **FIGURE 2**). The major conurbation is in the north-east region. It is often called BosNYWash because it covers the area from Boston in the north, through New York to Washington in the south. This region is home to over 50 million people (17% of the US population) and accounts for 20 per cent of the gross domestic product (GDP) of the United States.

FIGURE 2 Conurbations in the United States

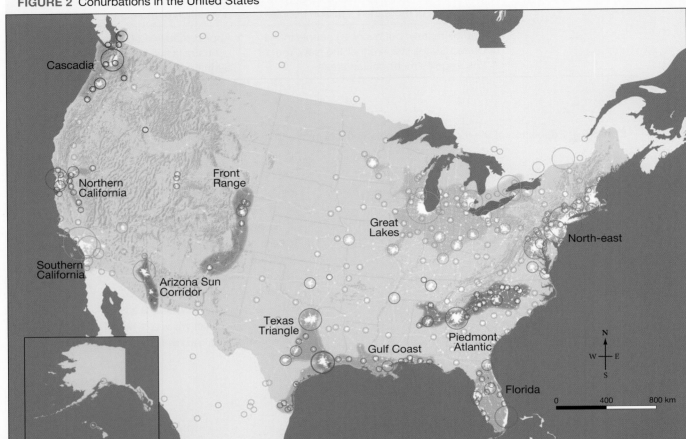

Source: Adapted with permission from Bernard Salt

Australia

Australia, on the other hand, has four conurbations (see **FIGURE 3**). One is in south-east Queensland, one joins Melbourne and Geelong, one is from Perth to Mandurah, and the Newcastle–Wollongong conurbation stretches for over 250 kilometres and is home to almost six million people.

7.5.4 Other consequences of urbanisation

Homelessness

According to the US National Alliance to End Homelessness, as of 2018 there were around 553 000 homeless people in the United States on a given night. This represents 17 people in every 10 000. Although the trend has been downwards from 2007–17, there was a slight rise in 2018. The five states with the highest homeless counts in 2018 were California (129 972), New York State (91 897), Florida (31 030), Texas (25 310) and Washington State (22 305).

In comparison, census data shows that the number of homeless people in Australia increased by more than 15 000 (14 per cent) over five years to 2016. According to the Australian Bureau of Statistics, 116 000 people were homeless on census night in 2016, representing 50 homeless people per 10 000. This was an increase of 13.7 per cent from the 2011 census.

FIGURE 3 Australia's population centres and conurbations

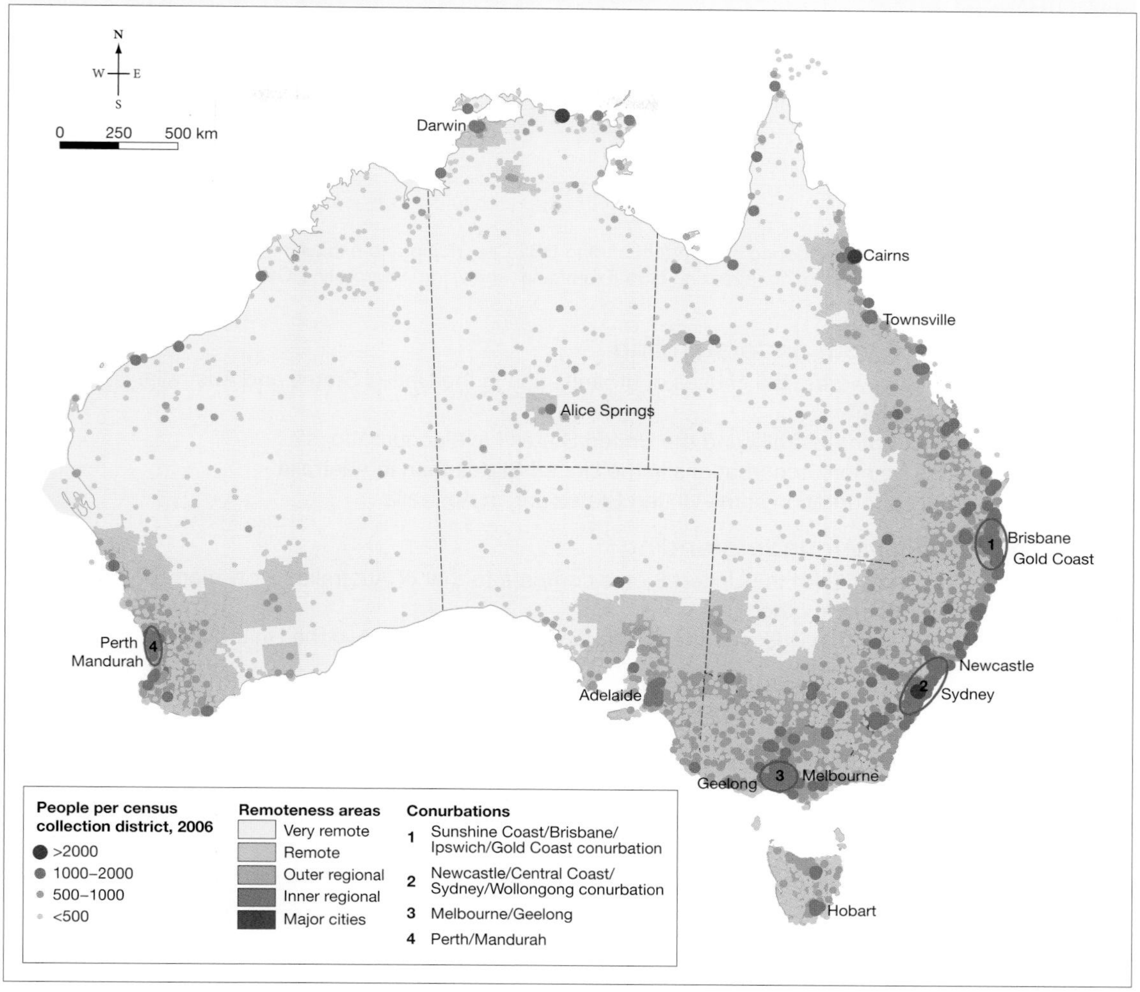

Darwin

Cairns

Townsville

Alice Springs

Brisbane
Gold Coast

Perth
Mandurah

Newcastle
Sydney

Adelaide

Geelong Melbourne

Hobart

People per census collection district, 2006	Remoteness areas	Conurbations
● >2000	☐ Very remote	1 Sunshine Coast/Brisbane/Ipswich/Gold Coast conurbation
● 1000–2000	☐ Remote	
● 500–1000	☐ Outer regional	2 Newcastle/Central Coast/Sydney/Wollongong conurbation
· <500	☐ Inner regional	3 Melbourne/Geelong
	☐ Major cities	4 Perth/Mandurah

Source: Australian Bureau of Statistics

Health issues

High population densities in urban areas make it easier for diseases to be transmitted, especially in poor neighbourhoods. The urban poor suffer health issues caused by reduced access to sanitation and hygiene facilities and health care.

Pollution

Air pollution from cars, industry and heating affects people who live in cities. A study in the United States showed that more than 3800 people die prematurely in the Los Angeles Basin and San Joaquin Valley region of southern California because of air pollution. Generally, Australia has a fairly high level of air quality. Cars and industry are the main factors influencing air quality in urban areas.

on Resources

🔷 **Interactivity** City folk (int-3117)

Explore more with myWorldAtlas

Deepen your understanding of this topic with related case studies and questions.
• Investigating Australian Curriculum topics > Year 8: Changing nations > **Urbanisation in Australia and the USA**

7.5 EXERCISES
Geographical skills key: GS1 Remembering and understanding **GS2** Describing and explaining **GS3** Comparing and contrasting **GS4** Classifying, organising, constructing **GS5** Examining, analysing, interpreting **GS6** Evaluating, predicting, proposing

7.5 Exercise 1: Check your understanding
1. **GS2** Explain, in your own words, the causes of urbanisation in the United States and Australia.
2. **GS1** What is a conurbation?
3. **GS2** Why do you think both Australia and the United States have conurbations?
4. **GS3** Why might there be more conurbations in the United States than in Australia?
5. **GS1** Name the largest conurbation in the United States and in Australia.

7.5 Exercise 2: Apply your understanding
1. **GS3** How does the population of the United States compare to that of Australia? How many times larger (approximately) is one than the other?
2. **GS2** Refer to **FIGURES 2** and **3**. Describe the distribution of the population in the United States and in Australia.
3. **GS3** Refer to **TABLE 1**.
   (a) Compare the *scale* of urbanisation in the United States and in Australia.
   (b) Compare the numbers of people living in large cities in the United States and in Australia.
4. **GS3** Refer to **FIGURE 1**.
   (a) Compare the size of the 10 largest cities in the United States and in Australia.
   (b) What might explain the differences you noticed?
5. **GS2** Apart from conurbations, what are three consequences of urbanisation?

Try these questions in learnON for instant, corrective feedback. Go to www.jacplus.com.au.

7.6 Effects of international migration on Australia

7.6.1 Why have people migrated to Australia?
Australia is a land of **migrants**. In a way all non-Indigenous Australian people are migrants — at some stage in the past, our ancestors came to this country to live. In 2016, nearly half of Australia's population was born overseas or had at least one parent who was born overseas.

Since the earliest times, people have moved from one part of the world to another in search of places to live. Migrants have come to Australia for many reasons (see **FIGURE 1**).

7.6.2 Where have our migrants come from?
Between 1851 and 1861 over 600 000 people came to Australia. While the majority were from Britain and Ireland, 60 000 came from Continental Europe, 42 000 from China, 10 000 from the United States and just over 5000 from New Zealand and the South Pacific. However, since 1975, the country has attracted more immigrants from Asia (see **FIGURE 3** and **TABLE 2**). Despite this, the most common ancestries today are still English, Australian, Irish, Scottish and Italian (see **TABLE 1**).

FIGURE 1 Reasons for immigration to Australia

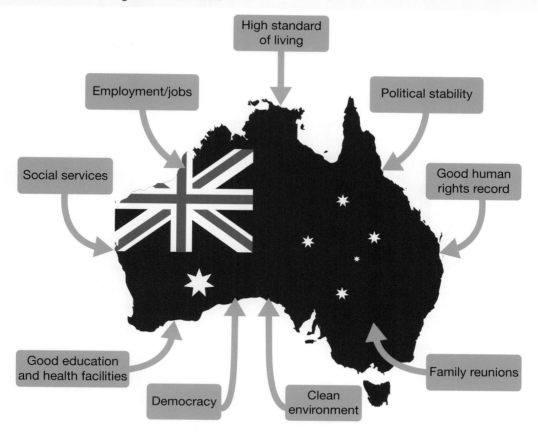

FIGURE 2 Origin of Australia's migrants, 1949–1959

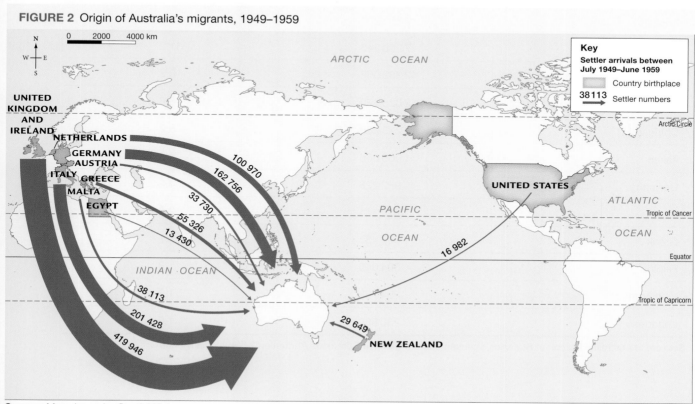

Source: Map drawn by Spatial Vision

FIGURE 3 Settler arrivals by country of birth according to the 2016 census

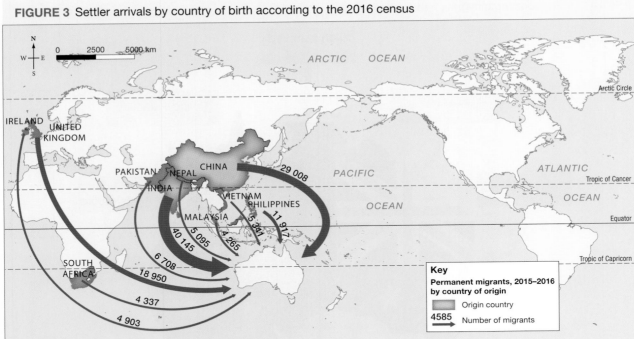

Key
Permanent migrants, 2015–2016
by country of origin

Origin country

4585   Number of migrants

Source: Department of Immigration and Border Protection

TABLE 1 Ancestry by birthplace of parents, 2016

Ancestry (top responses)	Number of Australians	Percentage
English	7 852 224	33.6
Australian	7 298 243	31.2
Irish	2 388 058	10.2
Scottish	2 023 470	8.6
Chinese	1 213 903	5.2
Italian	1 000 006	4.3
German	982 226	4.2
Indian	619 164	2.6
Greek	397 431	1.7
Filipino	304 015	1.3
Vietnamese	294 798	1.3
Lebanese	230 869	1.0

Source: © Australian Bureau of Statistics, licensed under a Creative Commons Attribution 2.5 Australia licence

Where have our migrants settled?

When they arrive, migrants tend to live in capital cities because of the greater availability of jobs and to be near family members, friends and people from the same country (see **TABLE 2**). In 2016, 83 per cent of the overseas-born population in Australia lived in capital cities. About one-third of the population in our large cities was born overseas.

Overseas-born migrants who arrived in the past 20 years are more likely to live in a capital city than those who arrived before 1992 (85 per cent compared to 79 per cent).

Migrants from certain countries tend to be attracted to certain Australian states or territories more than others (see **TABLE 3**).

TABLE 2 Top 10 birthplaces of Australians, 2016

Country of birth	Number of people	Percentage of state population	Percentage of state population living in capital city
United Kingdom	1 087 749	4.6%	5.0
New Zealand	518 466	2.2%	2.3
China	509 563	2.2%	3.1
India	455 388	1.9%	2.7
Philippines	232 397	1.0%	1.2
Vietnam	219 349	0.9%	1.4
Italy	174 051	0.7%	1.0
South Africa	162 450	0.7%	0.8
Malaysia	138 371	0.6%	0.8
Sri Lanka	109 841	0.5%	0.7

Source: © Australian Bureau of Statistics

TABLE 3 Top four countries of birth by state or territory ('000), 2016

ACT	NSW	NT	Qld	SA	TAS	VIC	WA
England (13.3)	China (256.0)	England (6.7)	New Zealand (200.4)	England (103.7)	England (20.5)	England (192.7)	England 213.9
China (11.9)	England (250.7)	Philippines (7.0)	England (219.9)	India (29.0)	New Zealand (5.4)	India (182.8)	New Zealand (87.4)
India (10.9)	India (153.8)	New Zealand (5.6)	India (53.1)	China (26.8)	China (3.3)	China (176.6)	India (53.4)
New Zealand (5.0)	New Zealand (127.9)	India (4.2)	China (51.6)	Italy (20.2)	India (2.1)	New Zealand (102.7)	Philippines (33.4)

Source: © Australian Bureau of Statistics

For example:

1. In 2016, Western Australia had the highest proportion of residents that were born in England of any state or territory (8.4%), more than twice the Australian proportion of 4.1 per cent.
2. Western Australia recorded the highest proportion of the population born overseas at 35 per cent (895 400 persons).
3. Victoria recorded the second highest proportion with 30.7 per cent of its residents born overseas (1 892 500 persons).
4. Queensland had the highest proportion of the population that were born in New Zealand (4.5%).
5. New South Wales had a higher proportion of residents born in China (3.3%) and South Korea (0.8%) than any other state or territory.
6. Victoria had the highest proportions of residents born in India (3.0%), Vietnam (1.5%), Italy (1.3%), Sri Lanka (1.0%) and Greece (0.9%).
7. The Northern Territory had the highest proportion of people born in the Philippines (2.8%).

Not only have immigrants tended to settle in larger cities, they have settled in particular suburbs and regions within the capital cities. Many migrants have settled in inner Sydney, for example, and especially in western Sydney suburbs (see **FIGURE 4**).

FIGURE 4 Distribution of new overseas migrants to Sydney, 2017

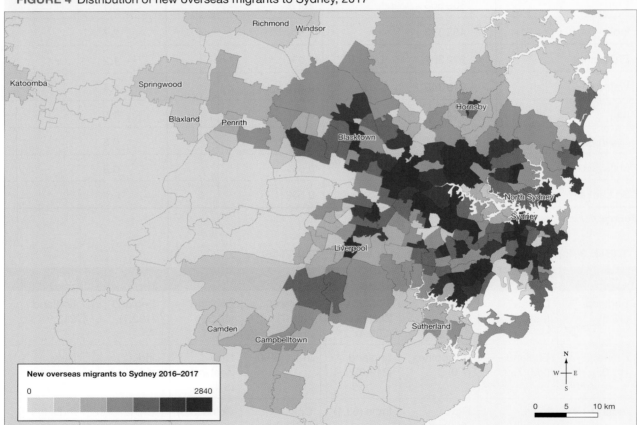

Source: The Sydney Morning Herald

7.6.3 Effects of international migration

Social effects

Migration has helped increase Australia's population. The increase in population from only seven million at the end of World War II to more than triple that now is caused by both the arrival of migrants and increased birth rates since then (see **FIGURE 5**).

Migrants to Australia have contributed to our society, culture and prosperity. Many communities hold festivals and cultural events where we can all share and enjoy the foods, languages, music, customs, art and dance.

Australian society is made up of people from many different backgrounds and origins. We have come from more than 200 countries to live here. Therefore, we are a very multicultural society — one which needs to respect and support each other's differences, and the rights of everyone to have their own culture, language and religion.

FIGURE 5 Australia's population growth, 1900–2017

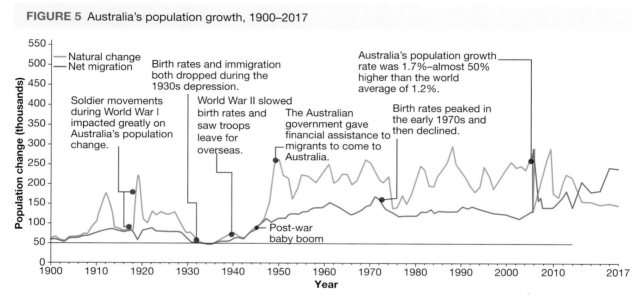

Economic effects

An increased population also means a greater demand for goods and services, which stimulates the economy. Migrants need food, housing, education and health services, and their taxes and spending allows businesses to expand. Apart from labour and capital (money), migrants also bring many skills to Australia (see **FIGURE 6**).

FIGURE 6 Types of migrants to Australia, 1991–2017

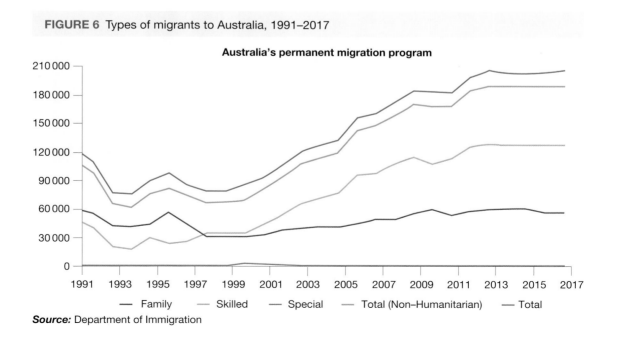

Source: Department of Immigration

Migrants generate more in taxes than they consume in benefits and government goods and services. As a result, migrants as a whole contribute more financially than they take from society.

Environmental effects

In the past, people argued that immigrants put pressures on Australia's environment and resources by increasing our population and the need for water, energy and other requirements. However, today many people believe that Australia's environmental problems are not caused by migration and population increase, but by inadequate planning and management.

Explore more with my**World**Atlas

Deepen your understanding of this topic with related case studies and questions.
• Investigating Australian Curriculum topics > Year 8: Changing nations > **International migration and Australian cities**

7.6.4 The future

Since 1995, the Australian government has been working to encourage new migrants to settle in regional and rural Australia. The Regional Sponsored Migration Scheme (RSMS) allows employers in areas of Australia that are regional, remote or have low population growth to sponsor employees to work with them in those regions (see **FIGURE 7**). This takes the pressure off large cities and also provides regional employers with skilled workers. As we have seen, it has always been the case that most immigrants settle first in our cities, especially the state capitals. In 2017–18, 101 255 migrants arrived in Australia and of these, 6637 settled in regional Australia. There are many regional locations that want to attract migrants.

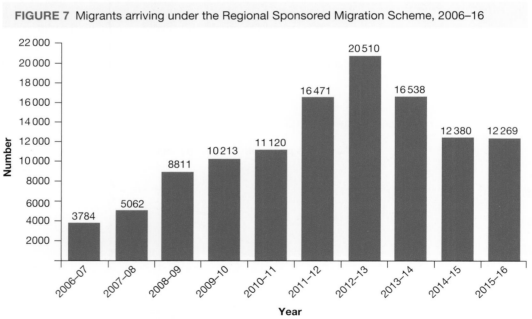

FIGURE 7 Migrants arriving under the Regional Sponsored Migration Scheme, 2006–16

Source: Australian Government, Department of Immigration and Border Protection, *2015–16 Migration Programme Report*

7.6 EXERCISES

Geographical skills key: GS1 Remembering and understanding **GS2** Describing and explaining **GS3** Comparing and contrasting **GS4** Classifying, organising, constructing **GS5** Examining, analysing, interpreting **GS6** Evaluating, predicting, proposing

7.6 Exercise 1: Check your understanding

1. **GS2** Using statistics, describe how Australia is truly a land of migrants.
2. **GS2** Refer to **FIGURES** 2 and 3. Describe how the origins of our migrants have **changed** since 1949.
3. **GS2** Refer to **FIGURE 7**. Describe how the number of migrants coming into Australia under the Regional Sponsored Migration Scheme has **changed** between 2006 and 2016.
4. **GS2** Refer to **FIGURE 5**. Describe how important migration has been in terms of Australia's population growth.
5. **GS5** Look at **FIGURE 6**. Which two categories provide the greatest number of migrants to Australia?

7.6 Exercise 2: Apply your understanding

1. **GS5** Refer to **TABLE 3** and **FIGURE 4**. Describe how the distribution of the areas of settlement by migrants varies within Australia.
2. **GS6** What do you consider to be the main reasons for why people would migrate to Australia?
3. **GS6** What do you believe are the two main benefits of migration to Australia? Give reasons for your answer.
4. **GS5** Study **FIGURE 5**. What impact did World War II have on Australia's birth rate and why?
5. **GS5** Study **FIGURE 5**. In which year was Australia's birth rate almost 50 per cent more than the world average? Suggest a reason for this.

Try these questions in learnON for instant, corrective feedback. Go to www.jacplus.com.au.

7.7 SkillBuilder: Creating and reading pictographs

What is a pictograph?

A pictograph is a graph drawn using pictures to represent numbers, instead of bars or dots that are traditionally used on graphs. A pictograph is a simple way of representing data and conveying information quickly and efficiently in a different format.

Select your learnON format to access:

- an overview of the skill and its application in Geography (Tell me)
- a video and a step-by-step process to explain the skill (Show me)
- an activity and interactivity for you to practise the skill (Let me do it)
- questions to consolidate your understanding of the skill.

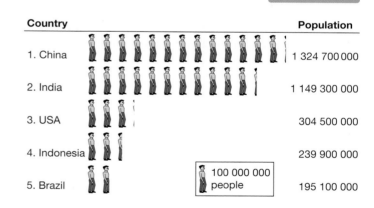

Country		Population
1. China		1 324 700 000
2. India		1 149 300 000
3. USA		304 500 000
4. Indonesia		239 900 000
5. Brazil		195 100 000

100 000 000 people

 Resources

Video eLesson SkillBuilder: Creating and reading pictographs (eles-1659)

Interactivity SkillBuilder: Creating and reading pictographs (int-3155)

7.8 People on the move in Australia and China

7.8.1 What makes Australians move?

In the United States, it is common for young people to leave home and travel to a university in another state or on the opposite side of the country. This is less common in Australia.

People move for many reasons. The average Australian will live in 11 houses during their lifetime — this means that many people will live in more. You may move to live in a larger house, or a smaller house as your family size or income changes. On retirement you may want to live near the mountains or the sea.

Thirty-nine per cent of Australians changed the place where they lived in the five years between 2006 and 2011. Most of the moves were limited to local areas especially within capital cities. About 4.4 per cent of moves involved a change of state or territory.

The major movements of Australians since 1788 are shown in **FIGURE 1**. The Great Australian Divide separates Australia into two regions, known as the Heartland and the Frontier. The Heartland is home to over 19 million people who live in a modern, urbanised, industrial state. The Frontier is a sparsely populated region of around three million people who live in a place that is remote but rich in resources.

Sea change or tree change

The population movement caused by 'sea change' or 'tree change' — a move from an urban environment to a rural location — is a national issue affecting coastal and forested mountain communities in every state in Australia. The movement involves people who are searching for a more peaceful or meaningful existence, who want to know their neighbours and have plenty of time to relax. Local communities in high-growth coastal and mountain areas often cannot afford the services and increased infrastructure, such as roads, water and sewerage, that a larger population requires. Geelong, Wollongong, Cairns and the Gold Coast are all popular places for sea changers to settle.

Not every sea changer loves their new life, and many return to the city. Factors such as distance from family, friends, cultural activities and various professional or health services may pull people back to their previous city residences.

Fly-in, fly-out workers

Employment opportunities have grown within the mining industry in places such as the Pilbara. However, local towns do not have the infrastructure, such as water, power and other services, to support a large population increase. Rental payments for homes can be as high as $3000 per week. One way to attract workers to these regions is to have a **fly-in, fly-out (FIFO)** workforce. FIFO workers are not actually 'settlers', as they choose not to live where they work. Some mine workers from the Pilbara live in Perth or even Bali, and commute to their workplace on a weekly, fortnightly or longer-term basis. The permanent residents of these remote towns are uneasy with the effects of the FIFO workforce because they change the nature of the town but choose not to make it their home. By not living locally, their wages leave the region and are not invested in local businesses and services.

Seasonal agricultural workers

Many jobs in rural areas are seasonal — for example, the picking and pruning of grapes and fruit trees requires a large workforce for only a few months each year. Many children born in rural areas leave their homes and move to the city for education, employment or a more exciting lifestyle than the one they knew in the country. This means that there are not enough agricultural workers to cover the seasonal activities.

FIGURE 1 Australia's moving population

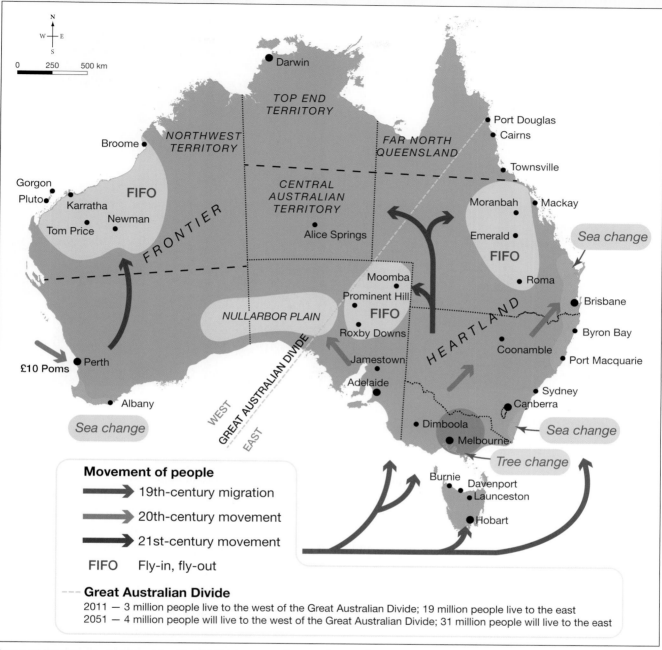

Source: Map drawn by MAPgraphics Pty Ltd, Brisbane; most recent data available at time of publishing

Backpackers plus people from Asia and the Pacific Islands on short-term work visas often provide the seasonal workforce in these regions. Country towns such as Robinvale in northern Victoria now have Asian grocery stores, an Asian bakery and a shop selling Tongan canned goods, providing the seasonal farm workforce with a taste of home. Robinvale has many people from different nationalities living as both permanent residents and seasonal workers. These include people from Italy, Tonga, Vietnam, Malaysia, New Zealand, China and Greece.

FIGURE 2 Newman, a mining town in the Pilbara region of Western Australia, provides accommodation for FIFO miners

7.8.2 Reasons for rural–urban migration in China

China has been experiencing a changing population distribution. The country's urban population became larger than that of rural areas for the first time in its history in 2012, as rural people moved to towns and cities to seek better living standards. China has become the world's largest urban nation.

Chinese labourers from the provinces have been moving to coastal cities in search of job opportunities, following reforms in 1978 that opened up China to foreign investment. Until then, rural–urban migration was strictly forbidden in China. Since then, more than 150 million peasants have migrated from the inner provinces to cities, mainly on the east coast. About half of rural migrants moved across provinces. This is the largest migration wave in human history (see **FIGURE 3**).

Pull factors

Migrants from rural areas are attracted to urban regions largely for economic reasons — a higher income is achievable in a city (see **FIGURE 3**). The average income of rural residents is about one-fifth that of urban residents on the east coast of China. Social factors are also important, with more opportunities for career development being available in cities; many people also desire a more modern urban lifestyle, with the benefits brought about by access to improved infrastructure and technology.

Push factors

Increasing agricultural productivity since the late 1970s has resulted in fewer labourers being needed on farms and thus a huge surplus of rural workers. These people have been forced to move to more urban areas in order to find employment. Agricultural production has meanwhile become less profitable, so workers have again been driven to cities to try to improve their economic situations (see **FIGURE 4**).

Political factors are also influential. China's central planners have encouraged local leaders in poor regions to encourage people to move to the cities. Their slogan was 'the migration of one person frees the entire household from poverty'.

FIGURE 3 People from Chinese inland provinces with lower wages and Human Development Index (HDI) values have moved to cities and provinces with higher HDIs and incomes.

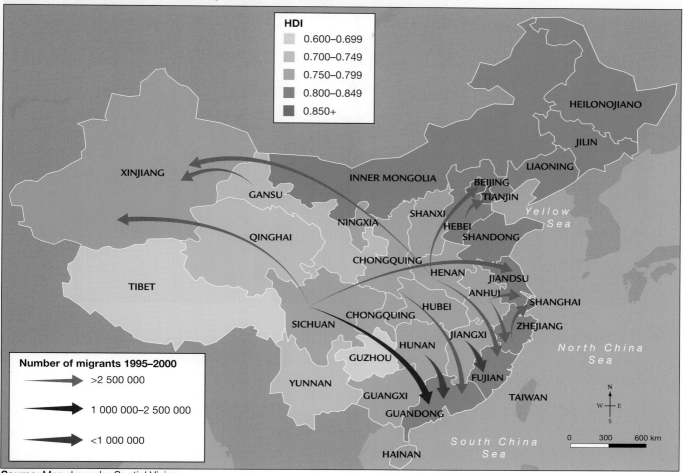

HDI
- 0.600–0.699
- 0.700–0.749
- 0.750–0.799
- 0.800–0.849
- 0.850+

Number of migrants 1995–2000
- >2 500 000
- 1 000 000–2 500 000
- <1 000 000

Source: Map drawn by Spatial Vision

FIGURE 4 A dramatic rural–urban migration shift has been occurring in China.

FIGURE 5 In 2017, Shanghai's population was estimated to be 24.21 million.

7.8.3 Consequences of rural–urban migration

- China's urban population rose from around 170 million people in 1978 to 540 million in 2004, and then to nearly 839 million in 2018.
- In 1949, 89 per cent of people lived in rural areas; by 1979 this figure had dropped to 81 per cent. In 2018 it was 59.3 per cent.
- It is expected that by 2050, only 25 per cent of China's population will be living in rural areas, while the number of city-dwellers will reach 940 million people.
- Some people predict that by 2025, China will have 19 super-cities with an average population of 25 million people each.
- Labourers from rural regions working in cities have to leave their families for months at a time or more.
- Tens of millions of people are classified as rural dwellers, even though they spend most or all of their time working in the cities. These people are denied access to social services, including subsidised housing, income support and education for their children.
- A shift to an increased urban population results in reduced population pressures on the land.
- Up to 40 per cent of rural income comes from urban workers sending money to their families at home.

DISCUSS

If 'a shift to an increased urban population results in reduced population pressures on the land', discuss what pressures might be added to urban areas. **[Critical and Creative Thinking Capability]**

 Resources

 Interactivity   Urban/rural China (int-3116)

Weblink   China's urban growth

Google Earth   Shanghai

7.8 INQUIRY ACTIVITIES

1. Use internet sources (such as the **China's urban growth** weblink in the Resources tab) to respond to the following:
   (a) Describe population *changes* in the various cities in China.
   (b) 'The largest population growth has occurred in cities on China's coastline.' How true is this statement? Explain your answer using figures from the website you used. **Examining, analysing, interpreting**
2. Creatively (in graphic or diagrammatic form) present some of the dramatic statistics in this subtopic to inform others of the *scale* of the *changes* happening to the distribution of China's population.
   Classifying, organising, constructing

7.8 EXERCISES

Geographical skills key: GS1 Remembering and understanding **GS2** Describing and explaining **GS3** Comparing and contrasting **GS4** Classifying, organising, constructing **GS5** Examining, analysing, interpreting **GS6** Evaluating, predicting, proposing

7.8 Exercise 1: Check your understanding

1. **GS1** What does FIFO mean?
2. **GS1** What is the difference between a *tree changer* and a *sea changer*?
3. **GS2** List the positive and negative factors of making a tree *change* or sea *change* as a:
   (a) family with young children
   (b) retired couple.
4. **GS1** How has the percentage of people living in China's rural areas *changed* since 1949? What is this number expected to be in the future?
5. **GS2** Describe the main *changes* that have occurred within China's urban population since 1978.

7.8 Exercise 2: Apply your understanding

1. **GS5** Look carefully at **FIGURE 1** and explain how the gap between Australia's east and west is predicted to alter over the next 40 years.
2. **GS6** A more recent population migration is towards high-rise apartment living in the centre of major cities. How might this trend impact on these new residents and the *sustainability* of the *environment* their migration is creating? Use examples to justify your stance.
3. **GS2** Explain in your own words the main reasons for the dramatic *change* in China's population distribution.
4. **GS4** Classify each of the various consequences of this *change* as positive or negative.
5. **GS2** Refer to **FIGURE 1**. Explain the difference between Australia's Heartland and its Frontier.

Try these questions in learnON for instant, corrective feedback. Go to www.jacplus.com.au.

7.9 SkillBuilder: Comparing population profiles

What is a population profile?

A population profile, sometimes called a population pyramid, is a bar graph that provides information about the age and gender of a population. The shape of the population profile tells us about a particular population. Comparing population profiles of different places helps us try to understand how and why they may be similar or different.

Select your learnON format to access:

- an overview of the skill and its application in Geography (Tell me)
- a video and a step-by-step process to explain the skill (Show me)
- an activity and interactivity for you to practise the skill (Let me do it)
- questions to consolidate your understanding of the skill.

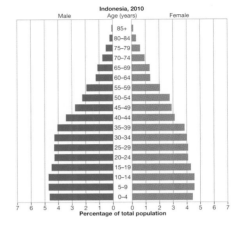

on Resources

Video eLesson	SkillBuilder: Comparing population profiles	(eles-1704)
Interactivity	SkillBuilder: Comparing population profiles	(int-3284)
Weblink	Population pyramid	

7.10 Thinking Big research project: Multicultural Australia photo essay

SCENARIO

Australia is celebrating a new national holiday – Multicultural Australia Day – to acknowledge the fact that Australia is made up of people from many backgrounds and origins. You are entering the inaugural photo essay competition, which aims to show aspects of Australia's rich multicultural heritage.

Select your learnON format to access:

- the full project scenario
- details of the project task
- resources to guide your project work
- an assessment rubric.

on Resources

 projectsPLUS Thinking Big research project: Multicultural Australia photo essay (pro-0173)

7.11 Review

7.11.1 Key knowledge summary

Use this dot point summary to review the content covered in this topic.

7.11.2 Reflection

Reflect on your learning using the activities and resources provided.

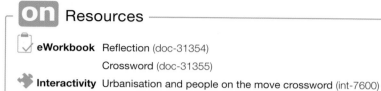 Resources

eWorkbook Reflection (doc-31354)

Crossword (doc-31355)

Interactivity Urbanisation and people on the move crossword (int-7600)

KEY TERMS

country the area of land, river and sea that is the traditional land of each Aboriginal language group or community; the place where they live

ecological footprint the amount of productive land needed on average by each person in a selected area for food, water, transport, housing and waste management

fly-in, fly-out (FIFO) a system in which workers fly to work, in places such as remote mines, and after a week or more fly back to their home elsewhere

geographical factors reasons for spatial patterns, including patterns noticeable in the landscape, topography, climate and population

indigenous native to or belonging to a particular region or country

migrant a person who leaves their own country to go and live in another

population density the number of people living within one square kilometre of land; it identifies the intensity of land use or how crowded a place is

population distribution the pattern of where people live; population distribution is not even — cities have high population densities and remote places such as deserts usually have low population densities

sea change movement of people from major cities to live near the coast to achieve a change of lifestyle

tree change movement of people from major cities to live near the forest to achieve a change of lifestyle

urban relating to a city or town; the definition of an urban area varies from one country to another depending on population size and density

urbanisation the growth and expansion of urban areas

urban sprawl the spreading of urban areas into surrounding rural areas to accommodate an expanding population

8 Our changing urban world

8.1 Overview

From cities to megacities to megaregions, why are the world's urban areas on the rise?

8.1.1 Introduction

In 2008, for the first time in history, the majority of the world's population lived and worked in towns and cities. This urban population is projected to continue growing in the future. The fast pace and unplanned nature of this growth has seen the development of megacities. However, along with the opportunities provided by the megacities come many problems. It is a challenge to create sustainable urban environments that meet the needs of the people living in these places.

Resources

☑ **eWorkbook**   Customisable worksheets for this topic

🎞 **Video eLesson** Megacities and megaregions (eles-1629)

LEARNING SEQUENCE

8.1 Overview
8.2 Urban areas and their effects on people
8.3 **SkillBuilder:** Describing photographs `online only`
8.4 Cities and megacities of the world
8.5 Causes and effects of Indonesia's urban growth
8.6 **SkillBuilder:** Creating and reading compound bar graphs `online only`
8.7 Characteristics of cities around the world
8.8 Creating sustainable cities
8.9 Sustainable cities in Australia
8.10 **SkillBuilder:** Constructing a basic sketch map `online only`
8.11 **Thinking Big research project:** One day in Jakarta, one day in New York City `online only`
8.12 **Review** `online only`

To access a pre-test and starter questions and receive immediate, **corrective feedback** and **sample responses** to every question, select your learnON format at www.jacplus.com.au.

8.2 Urban areas and their effects on people

8.2.1 Why people move to urban areas

There are many and varied reasons for people migrating to urban locations. These reasons are usually a combination of push and pull factors. Some people are 'pushed' from rural to urban areas within their own country. Others will travel from other countries to urban areas, 'pulled' by better opportunities.

Push factors

Geographical inequality is mostly responsible for the **migration** of people from rural to urban areas. **Push factors** that drive people towards cities usually involve a decline in living conditions in the rural area in which the people live. There are various situations that can cause this, including a decrease in the quality of agricultural land (caused by factors such as prolonged drought, erosion or desertification); poverty; lack of medical services or educational opportunities; war; famine from lack of food and/or crop failure; and natural disasters.

Pull factors

Pull factors refer to the attractions of urban areas that make people want to move there. Urbanisation in any country generally begins when enough businesses are established in the cities to provide many new jobs. Pull factors include job opportunities; better housing and infrastructure; political or religious freedom; improved education and healthcare; activities and enjoyment of public facilities; and family links.

FIGURE 1 Examples of push factors include lack of medical services, war, crop failure, prolonged drought and desertification, famine, poverty and lack of educational opportunities.

FIGURE 2 Examples of pull factors include religious tolerance, improved healthcare, job opportunities, family links, better housing and infrastructure, political freedom and better educational opportunities.

8.2.2 CASE STUDY: Migration due to climate change in India and Bangladesh

Climate change has resulted in higher temperatures, more extreme weather, rising sea levels, flooding and increased cyclonic activity in the Bay of Bengal and the Arabian Sea. These changes have affected the environment in many places in Bangladesh and India.

Bangladesh is a low-lying country with a dense population. The population in many regions rely on farming for their livelihood. Rising sea levels have introduced salt water into rice fields and reduced food production, income and job opportunities. Along with the attraction of jobs in construction in India, this has resulted in international migration from Bangladesh to India.

India also experiences climate change issues — flooding, erosion and landslides and areas of drought. The stresses caused by these issues have had an effect on millions of people in this region and has led to internal migration, particularly from rural to urban areas. The population density in 2018 in Mumbai was over 28 000 people per square kilometre; in Delhi it was 12 600 people per square kilometre. This movement of people has also increased tensions and conflict between ethnic groups, including over land rights.

FIGURE 3 Migration flows in India

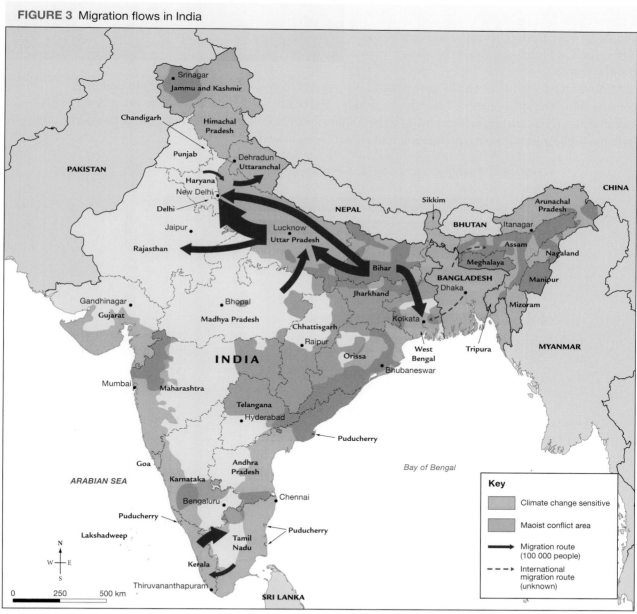

Source: Bhattacharyya and Werz

8.2.3 Which cities attract workers?

Taxi drivers, construction workers, teachers, nurses, house cleaners, accountants, nannies — there are many job opportunities for both skilled and unskilled workers that attract people to cities. These people may come from a different area within a country or across borders from different countries.

'Gateway cities' are cities in the world that are arrival points for many migrant workers. These cities are large enough to provide many different jobs and are therefore attractive to people moving from other regions. Some cities, such as Dubai, are reliant on their foreign workers.

FIGURE 4 These migrants are working in a fish-cleaning station in Dubai, United Arab Emirates.

More than two-thirds of Dubai's population is migrant labour, with many working in building construction. These labourers — mostly from India, Pakistan and Bangladesh — are often poorly paid, and live in migrant camps that can be up to two hours away from the work site. **FIGURE 5** shows the cities with the highest foreign-born population that attract foreign workers.

FIGURE 5 Foreign-born populations in gateway cities

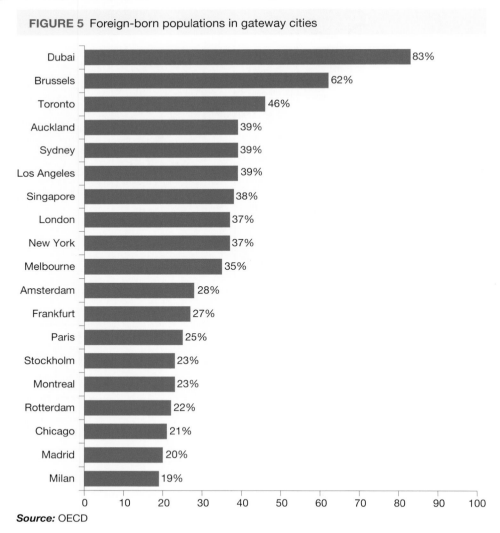

City	Percentage
Dubai	83%
Brussels	62%
Toronto	46%
Auckland	39%
Sydney	39%
Los Angeles	39%
Singapore	38%
London	37%
New York	37%
Melbourne	35%
Amsterdam	28%
Frankfurt	27%
Paris	25%
Stockholm	23%
Montreal	23%
Rotterdam	22%
Chicago	21%
Madrid	20%
Milan	19%

Source: OECD

8.2.4 CASE STUDY: Growth of cities in Africa

Africa now has a larger urban population than North America and has 25 of the world's fastest-growing large cities — the number of people living in cities in Africa is increasing by about one million every week. Some of Africa's cities are expected to grow by 85 per cent by 2025. By 2050, the urban population is expected to triple from 400 million people to 1.2 billion. Over half of the urban population in Angola, Chad, Madagascar, Malawi, Mozambique, Niger, Sierra Leone and Zambia is below the poverty line. In many other countries, including Burundi, Gambia, Kenya and Zimbabwe, 40–50 per cent of the population are living below the poverty line.

In most African cities, between 40 and 70 per cent of the population live in slums or squatter settlements. In cities such as Nairobi, Lagos, Cairo and Rwanda, 60–70 per cent of the population live in slum conditions, which occupy about five per cent of the land in the city.

FIGURE 6 The projected growth of African cities from 2010 to 2025

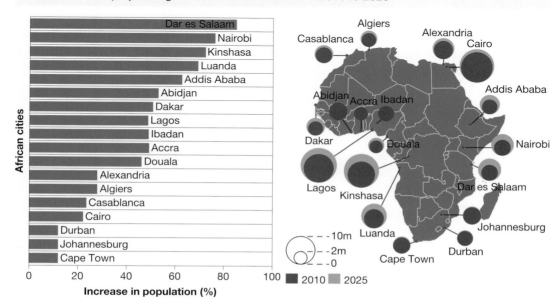

FIGURE 7 A slum in Nairobi

8.2.5 Regional differences

During the nineteenth and early twentieth centuries, urbanisation occurred because of migration and the growth of industries. New job opportunities in the cities attracted people from rural areas and migrants provided a cheap workforce for factories. At that time, death rates in cities were high because they were unhealthy places (with crowded living conditions, diseases and a lack of sanitation) and urban growth was slow. Workers often found it hard to find somewhere to live — it was not unusual for an entire family to be living in a single room. In many European cities (such as London) the number of deaths was higher than the number of births, and migrants provided most of the population growth.

It is a very different experience in developing countries today. Most urban growth results from natural increase; that is, people being born in cities, rather than migrating to cities. With the additional population increase caused by migration from rural areas in search of better jobs, many cities in Asia and Africa have exploded in size.

Cities can be great places and should not be viewed negatively. For example, people can more easily access basic services in urban areas than in rural areas so, although poverty may be present in urban environments, cities also offer an escape from poverty. Cultural activities are often enhanced in cities that attract migrants from many different areas — food and music are obvious examples. There also tends to be a greater tolerance of different migrant and racial groups living close together.

on Resources

⬧ **Interactivity** Urban push and pull factors (int-3118)

Explore more with my**World**Atlas

Deepen your understanding of this topic with related case studies and questions.
* Investigate additional topics > Urbanisation > **Mongolia**

8.2.6 How urban areas affect people's ways of life

Both small and large urban areas can provide people with positive and negative experiences.

Cities attract people to them with the opportunity of work and the possibility of better housing, education and health services. There is a strong interconnection between the wealth of a country and how urbanised it is. Generally, countries with a high **per capita income** tend to be more urbanised, while low-income countries are the least urbanised.

This happens because people grouped together create many chances to move out of poverty, generally because of increased work opportunities. There are often better support networks from governments and local councils. It is also cheaper to provide facilities such as housing, roads, public transport, hospitals and schools to a population concentrated into a smaller area.

8.2.7 Urban challenges

Rapid population growth in urban areas can result in problems such as poverty, unemployment, inadequate shelter, poor sanitation, dirty or depleted water supplies, air pollution, road congestion and overcrowded public transport.

Slums

In many developing countries, urban growth has resulted in unplanned settlements called **slums** (other terms used around the world include ghettos, favelas, shantytowns, bidonvilles and bustees). Almost 1 billion people live in slums worldwide.

The United Nations defines a slum as follows.

. . . one or a group of individuals living under the same roof in an urban area, lacking one or more of the following five amenities: (1) durable housing (a permanent structure providing protection from extreme climatic conditions); (2) sufficient living area (no more than three people sharing a room); (3) access to improved water (water that is sufficient, affordable and can be obtained without extreme effort); (4) access to improved sanitation facilities (a private toilet, or a public one shared with a reasonable number of people); and (5) secure tenure and protection against forced eviction.

Water and sanitation

Providing clean and safe water and sanitation in rural and urban areas is one of the Sustainable Development Goals (SDGs), Targets 6.1 and 6.2. Not all people have access to safe drinking water and to safely managed sanitation. In some countries, people still defecate in open areas. **FIGURES 8** and **9** show the differences in access to these facilities in rural and urban areas.

FIGURE 8 Urban and rural population with access to safe drinking water, 2015

FIGURE 9 Urban and rural population with access to safe sanitation, 2015

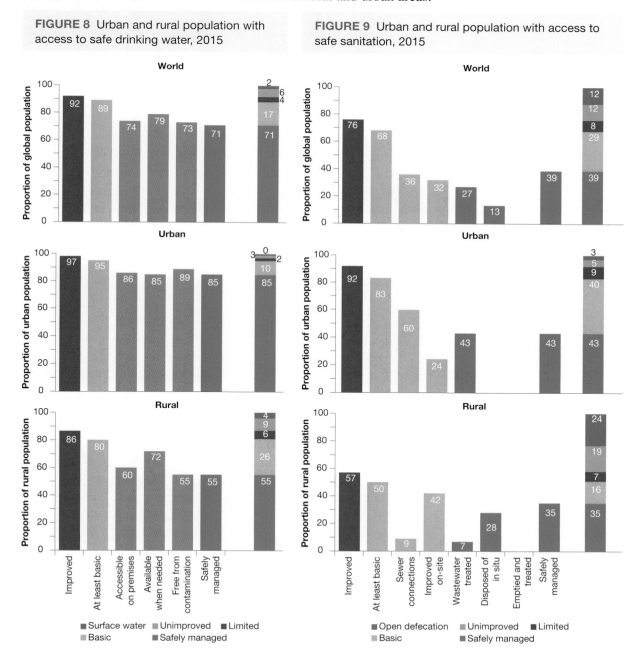

Transport and pollution

In cities that can't keep up with rapid population growth, traffic congestion and overcrowded public transport mean that many people must travel for hours to get to and from work (see **FIGURE 10**).

Pollution is also a problem that affects the health of people living in cities. Many cities have high levels of air pollution and some — including Mexico City, Buenos Aires, Beijing and Los Angeles — are famous for being so polluted.

According to the World Health Organization in 2016, 12 of the world's 25 cities with the worst air pollution were in India. Most of the pollution comes from the growing industrial sector and vehicle emissions.

FIGURE 10 Traffic congestion in Los Angeles, United States

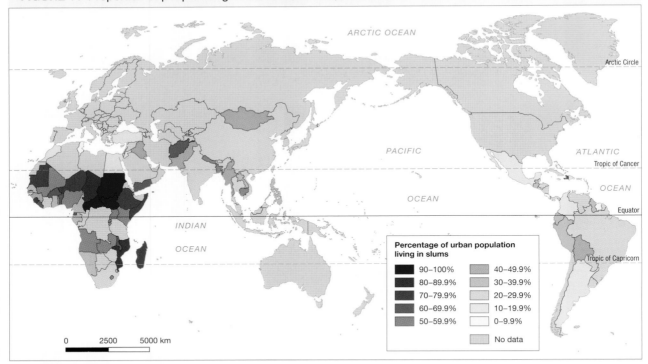

FIGURE 11 Proportion of people living in slums worldwide, 2015

Percentage of urban population living in slums

- 90–100%
- 80–89.9%
- 70–79.9%
- 60–69.9%
- 50–59.9%
- 40–49.9%
- 30–39.9%
- 20–29.9%
- 10–19.9%
- 0–9.9%
- No data

0   2500   5000 km

Source: World Bank Data

Explore more with my **World** Atlas

Deepen your understanding of this topic with related case studies and questions.
- Investigate additional topics > Urbanisation > **Mexico City**

8.2 INQUIRY ACTIVITIES

1. Use an atlas to locate all the gateway cities mentioned in **FIGURE 5**, and then mark their locations on a blank map of the world. When your map is complete, describe the distribution of the major gateway cities around the world.
Describing and explaining

2. Find out the population density of the capital city in your state or territory. How does it compare to that of Mumbai and New Delhi in 2018? List all the ways in which living in one of these Indian cities might be different to life in your local city.
Comparing and contrasting

3. Watch the video about urbanisation in **Mongolia** in *myWorld Atlas*.
   (a) List the push and pull factors that have caused people to move to Ulaanbaatar.
   (b) Describe the living conditions of these people. Do you think they are better or worse than their living conditions in rural areas? Justify your response.
Examining, analysing, interpreting

4. Why is it difficult in a country the size of Australia, with population concentrated on the coast, to provide services in outback areas? How would providing services be different in a country such as Luxembourg in Europe? Look at the size of Luxembourg in an atlas or by using Google Maps or Google Earth.
Comparing and contrasting

5. Use **FIGURE 2** in subtopic 8.4 and the *myWorld Atlas* statistical mapper to find the relationship between urbanisation and wealth. Give five examples of countries from different continents that are highly urbanised and wealthy, and five that are not urbanised and are poor. Do any countries not fit this pattern? Name them.
Classifying, organising, constructing

6. Conduct some research to find out which Australian city has the worst data for the two urban problems of transport and pollution.
Examining, analysing, interpreting

8.2 EXERCISES

Geographical skills key: GS1 Remembering and understanding **GS2** Describing and explaining **GS3** Comparing and contrasting **GS4** Classifying, organising, constructing **GS5** Examining, analysing, interpreting **GS6** Evaluating, predicting, proposing

8.2 Exercise 1: Check your understanding

1. **GS1** What are push factors? What are pull factors? Give two examples of each.
2. **GS1** Match each of the images in **FIGURES 1** and **2** with the push or pull factors listed in the captions.
3. **GS1** What is a gateway city? Why are people attracted to them?
4. **GS5** Which gateway city in **FIGURE 5** do you think would provide the greatest chance for foreign people to get work? Explain your answer.
5. **GS2** What is the difference between urban population increase from migration and from natural increase? Which of these is more likely to occur in a city located in a developing country? Why?
6. **GS1** What is a slum? Make a list of some other names for slums.
7. **GS1** Why are transport and pollution often problematic in large urban areas?
8. **GS2** Imagine you live in a poor rural village in India with no education or work. List the possible attractions of moving to an urban area.

8.2 Exercise 2: Apply your understanding

1. **GS5** Study **FIGURE 3** and the text in the case study 'Climate change and migration in India and Bangladesh'. Identify the push and pull factors that result in migration in India and Bangladesh.
2. **GS5** Study **FIGURE 6** and refer to an atlas map of Africa.
   (a) Name the three largest African cities in 2010 and the three predicted to be largest in 2025. In which countries are they located?
   (b) Describe the distribution of Africa's large cities. How many are inland? How many are on the coast? Which are located in the north, south-east and west of the continent? List the countries that do not have large cities.
   (c) What does it mean to live below the poverty line? Locate the cities in which more than half the population is living below the poverty line.
3. **GS4** Look at **FIGURE 7**. Draw a sketch of this scene and annotate it with geographical questions you would like answered about the **environment** and the people living there.
4. **GS6** What do you think is the future **sustainability** of the **place** shown in **FIGURE 7**, especially if the population of this city is going to increase?

▶

5. **GS5** Study **FIGURES 8** and **9**. Identify which of the following statements are true and which are false. Rewrite the false ones so they are true.
   (a) Providing safe access to drinking water to the world has been more successful than providing safe and adequate sanitation.
   (b) Most people in urban areas have access to safely managed drinking water.
   (c) Overall, a greater proportion of people in urban areas have access to improved drinking water.
   (d) Forty-three per cent of people in rural areas have unimproved or no access to sanitation.
   (e) Overall, people have better access to safe sanitation facilities than safe drinking water.
6. **GS5** Study **FIGURE 11**.
   (a) In which continent are the most urban slums found?
   (b) Name three countries in this continent with a very high proportion of people living in slums.
   (c) Describe the general pattern shown in the map.

Try these questions in learnON for instant, corrective feedback. Go to www.jacplus.com.au.

8.3 SkillBuilder: Describing photographs

What is meant by 'describing photographs'?

A description is a brief comment (up to a paragraph) on a photograph, identifying and communicating features from a geographic point of view. As geographers, we use our understanding of the world to interpret the image and tell others about the main features or information the photograph reveals.

Select your learnON format to access:

- an overview of the skill and its application in Geography (Tell me)
- a video and a step-by-step process to explain the skill (Show me)
- an activity and interactivity for you to practise the skill (Let me do it)
- questions to consolidate your understanding of the skill.

Resources

Video eLesson SkillBuilder: Describing photographs (eles-1660)

Interactivity SkillBuilder: Describing photographs (int-3156)

8.4 Cities and megacities of the world

8.4.1 Where are cities located?

How is a city different from other urban areas such as towns and villages? A city is a large and permanent settlement, and is usually quite complex in terms of transport, land use and **utilities** such as water, power and **sanitation**.

The image of the Earth at night (**FIGURE 1**) shows where lights are shining. The brightest areas on the map are the most urbanised, but might not be the most populated. If you compare this image with **FIGURE 2**, you can make some comparisons. For example, there are very bright lights in western Europe (Belgium, The Netherlands, France, Spain and Portugal, Germany, Switzerland, Italy and Austria) and yet more people living in China and India. Refer to your atlas to locate these countries.

The world's cities are generally located along or close to coastlines and transport routes. Some regions remain thinly populated and unlit. Antarctica is entirely dark. The interior jungles of Africa and South

America are mostly dark, but lights are beginning to appear there. Deserts in Africa, Arabia, Australia, Mongolia and the United States are poorly lit as well, although there are some lights along coastlines. Other dark areas include the forests of Canada and Russia, and the great mountains of the Himalayan region and Mongolia.

FIGURE 1 Satellite image of the Earth at night

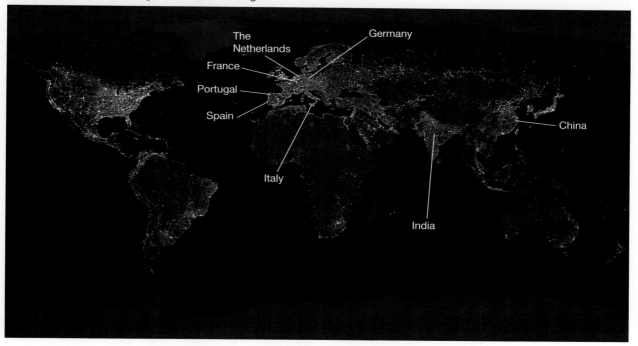

FIGURE 2 Population density and distribution of major cities in 2018, with selected city populations

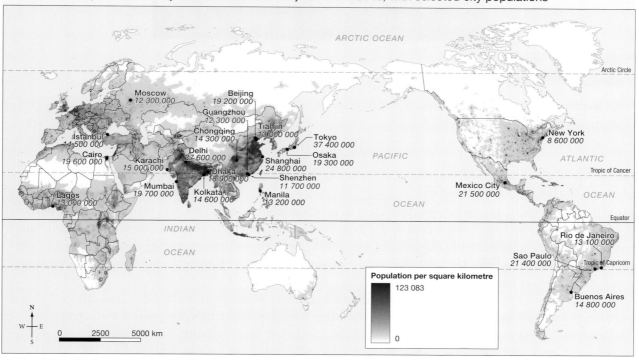

Source: United Nations, Department of Economic and Social Affairs, Population Division (2018). World Urbanization Prospects: The 2018 Revision.

FIGURE 3 Medellin, the second-largest city in Colombia, South America

 Resources

 Weblink   World City Populations

Google Earth Medellin, Colombia

8.4.2 What is a megacity?

Over the next century, urbanisation is predicted to increase at an even greater rate than it has in the past. Around the year 1900 only 15 per cent of the world's population lived in cities. At some time in 2007 this reached 50 per cent. In 2018, this figure was 55 per cent, with projections expecting that two-thirds of the population will live in cities by 2050. People are attracted to cities with huge populations, and increasingly these cities are becoming megacities.

A **megacity** is a city with more than 10 million inhabitants. When you consider that Australia's population was almost 25 million in 2018 — with over 5 million living in Sydney and 5 million in Melbourne — it is hard to imagine what it would be like to live in a megacity.

The number of megacities has grown over time. In 1950, only two cities in the world — Tokyo and New York — had a population above 10 million. By 1975 there were four; by 2000 there were 17, and in 2018 there were 33 megacities. By 2030, it is predicted that there will be 43 megacities in the world. Nineteen of these cities have a population greater than 15 million.

The distribution of megacities — that is, where they are located over space in the world — has also changed. In 1975, two megacities were located in the Americas and two in Asia. In 2014 more than half (15) of all megacities were located in Asia; and it is predicted that, in 2030, 23 of the 41 megacities will be located in Asia. There is also a change in terms of the wealth of countries that contain megacities, with the majority now located in developing countries. This is in contrast to the development of urbanisation, when North America and Europe were the focus of historic urban growth. By 2030, it is predicted that 23 megacities will exist in less developed countries.

FIGURE 4 Regional population distribution in different city sizes in 2018 and projected to 2030

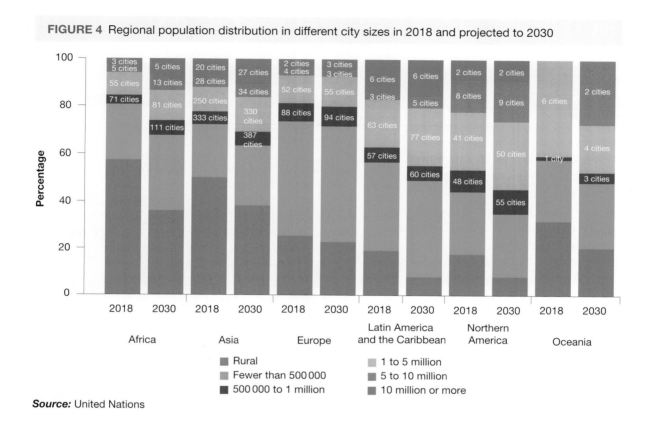

Source: United Nations

8.4.3 The never-ending city

In some parts of the world, megacities are merging to create **megaregions**. These regions are home to huge populations. Examples of megaregions include:

- Hong Kong–Shenzhen–Guangzhou in China, already home to 65 million people
- Kyoto–Osaka–Kobe, with a population of over 20 million in 2015.

Pearl River Delta (PRD)

This region is located in southern China on the South China Sea. The PRD is one of the fastest-growing regions in the world. There are five major cities — Hong Kong, Shenzhen, Dongguan, Foshan and Guangzhou and six smaller cities made up of Macau, Zhaoqing, Zhuhai, Jiangmen, Huizhou, and Zhongshan, which are linked by transport routes and provide great economic opportunities. Until 1979, Shenzhen was a fishing village. In 1980 the government declared the area to be a Special Economic Zone (SEZ), attracting businesses and investment from other countries. Since then, the area has undergone rapid urbanisation that has dramatically changed the landscape around the Pearl River Delta (see **FIGURE 5**).

FIGURE 5 Change in the Pearl River Delta between (a) 1988 and (b) 2014

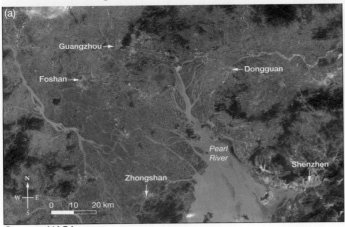

Source: NASA

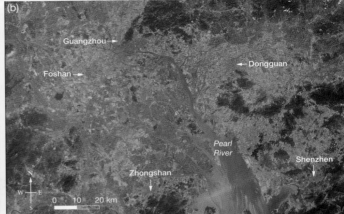

Source: NASA

In 1988, the rivers and streams flowed through a fertile region with rice paddies, wheat fields, orchards and fish ponds. The region was mostly rural, and the population of roughly 10 million distributed between rural areas and a few cities.

By 2014 these cities had grown quickly and merged into an interconnected megalopolis with a population of 42 million. When combining the population of Hong Kong, the total is around 65 million.

Megacity facts

- Over half the future growth in megacities will be within Asia.
- The 20 largest cities consume 80 per cent of the world's energy and produce 80 per cent of global greenhouse gas emissions.
- Slums in megacities are especially vulnerable to climate change, as they are often built on hazardous sites in high-risk locations.

FIGURE 6 The city of Shenzhen, in the Pearl River Delta, in the twenty-first century

 Resources

 Interactivity Megacity march (int-3119)

8.4 INQUIRY ACTIVITIES

1. Use a political map in your atlas and **FIGURE 2** to identify the following.
   (a) The Nile River
   (b) The Trans-Siberian railway from Moscow to Vladivostok
   (c) Highways linking cities in the western and eastern United States
   (d) The Himalayan mountain range

   Examining, analysing, interpreting

2. Go to the **World City Populations** weblink in the Resources tab. Work as a team of five and investigate the *change* in city population in different continents. Discuss which regions each member will investigate and record maps, data and graphs. Report your findings back to the group.

Classifying, organising, constructing

3. Use an atlas to locate the two megaregions mentioned in section 8.4.3. Why do these regions develop?

Describing and explaining

4. Research the 'dead zone' in the sea at the mouth of the Pearl River. What does this mean, and what is its cause?

Describing and explaining

5. Describe the *changes* that have occurred in the Pearl River Delta region. Find this *place* in an atlas and describe where it is in relation to the rest of China and to two other countries in Asia.

Describing and explaining

6. Work with another student to produce a Prezi or PowerPoint presentation or an animation showing the world's megacities in 2018 and 2030. Include images from the internet and data from **FIGURE 4**. You may like to choose appropriate music to accompany the presentation. **Classifying, organising, constructing**

7. After completing the 'Describing photographs' SkillBuilder in subtopic 8.3, complete the following questions about **FIGURE 3**.
   (a) Describe the foreground and background shown in the photograph.
   (b) List the natural and human characteristics shown in the photograph.
   (c) What does this photograph show about urban *environments*? How has the urban *environment changed* the natural *environment*?
   (d) How might the *changes* described in part (b) lead to an increased risk of erosion? (See topic 2 for information on erosion processes.)
   (e) Imagine that the population of this city continues to increase. Describe what might happen to the land in the future.
   (f) Do you think that all land surrounding cities should be able to be taken up by buildings? Why or why not?
   (g) Investigate the *place* where you live. Are there land-use zones that cannot be built upon, such as 'green wedges'? Where are they and why are they there? Do you think they should be protected from development? Justify your answer. **Examining, analysing, interpreting**

8.4 EXERCISES

Geographical skills key: GS1 Remembering and understanding **GS2** Describing and explaining **GS3** Comparing and contrasting **GS4** Classifying, organising, constructing **GS5** Examining, analysing, interpreting **GS6** Evaluating, predicting, proposing

8.4 Exercise 1: Check your understanding

1. **GS1** How is a city different from a town or a village?
2. **GS1** What do the bright lights in **FIGURE 1** show?
3. **GS1** What is a megacity? How many megacities were there in 2018?
4. **GS1** How many megacities are predicted by 2030?
5. **GS1** Name the first two megacities and the countries where they are located.
6. **GS1** What is a megaregion?

8.4 Exercise 2: Apply your understanding

1. **GS5** Study **FIGURES 1** and **2** and refer to a political map in your atlas. Which of the following statements are true and which are false? Rewrite the false statements to make them true.
   (a) Japan is a highly populated country with many cities.
   (b) The west coast of the United States is more densely populated than the east coast.
   (c) The Amazon rainforest does not have any settlements.
   (d) The eastern region of China has more cities than the western region.
   (e) The main city settlements in Australia are along the east coast.
   (f) The distribution of cities across Europe is uneven.

▶

2. **GS5** Refer to **FIGURE 4**. This is a compound bar graph (see SkillBuilder 8.6) showing the projected distribution of cities, including megacities, between 2018 and 2030. Study the graph and the statements below, identifying which are true and which are false. Rewrite the false ones to make them true.
   (a) Rural populations across all regions are declining.
   (b) The highest number of megacities are located in Latin America and the Caribbean.
   (c) In 2018 over half the population in Asia and Africa lived in rural areas.
   (d) In 2030 Europe will have the highest percentage of cities with fewer than 500 000 people.
3. **GS5** Describe the *changes* to the Pearl River Delta from 1988 to 2014.
4. **GS6** What impact will the *changes* identified in question 3 have on people and the *environment*?
5. **GS4** Study **FIGURE 4**. Create a list of regions from highest to lowest that will have the highest percentage of people living in rural areas in 2030.

Try these questions in learnON for instant, corrective feedback. Go to www.jacplus.com.au.

8.5 Causes and effects of Indonesia's urban growth

8.5.1 Indonesia's population

Many people do not realise that the fourth most populated country in the world is one of our nearest neighbours. Like many countries in Asia, Indonesia has experienced rapid urban growth, but this has occurred only relatively recently.

Indonesia's population of nearly 270 million people (2019) lives on a chain or cluster (an archipelago) of more than 18 000 islands (see **FIGURE 1**). However, its population is not evenly distributed. Only about 11 000 of the islands are actually inhabited. Sixty per cent of Indonesia's population is concentrated on only seven per cent of the total land area — on the island of Java.

FIGURE 1 Map of Indonesia

Source: Spatial Vision

Indonesia has changed from a rural to an urban society quite recently. In 1950, only 15.5 per cent of its population lived in urban areas. In 2018, this had increased to 55.3 per cent.

Like many countries in Asia, Indonesia has a high concentration of its urban population in a few large cities. In 1950, there was only one city that was home to more than one million people in Indonesia: Jakarta. That had increased to four cities by 1980, eight by 1990, 10 by 2000 and 14 by 2016. More than one-fifth of the Indonesian urban population now lives in the Jakarta metropolitan area (JMA).

8.5.2 Causes of urbanisation

More than one-third of Indonesia's urban population growth resulted from natural increase. It took until 1962 for Indonesia's population to reach 100 million people. However, it then took only until 1997 to reach 200 million. In the early 1970s, Indonesia's birth rate was very high — 5.6 children per woman. However, the growth rate has fallen dramatically from 2.3 per cent in 1970 to about 1.2 per cent in 2015. In 2018 there were nearly 5.5 million babies born in Indonesia — almost the equivalent of the population of Melbourne.

As few restrictions were placed on rural–urban migration, most of the migration movement consisted of the rural poor moving into cities and especially into slums, leaving their families behind in the villages. On top of this, in recent years about 20 000 foreigners per year have obtained work permits for Indonesia.

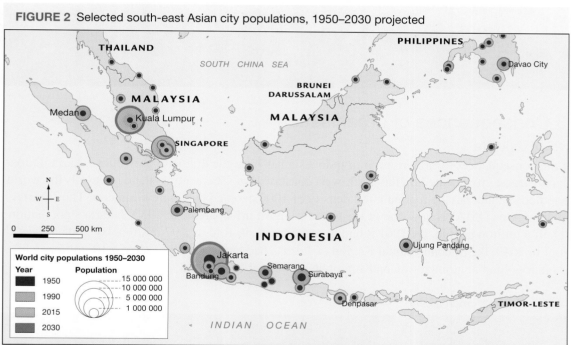

FIGURE 2 Selected south-east Asian city populations, 1950–2030 projected

Source: United Nations, Department of Economic and Social Affairs, Population Division (2014). World Urbanization Prospects: The 2014 Revision. CD-ROM Edition

Investment from within Indonesia and from other countries has tended to occur mainly in the large urban areas, because these areas can supply the workers, transport (by sea and over land), water and electricity that are needed by industry.

The first president of Indonesia wanted Jakarta to be like the world's great cities, such as Paris and New York, as well as a focus for other Indonesian cities. President Sukarno therefore built broad avenues, highways and electric railway lines, luxurious housing estates, high-rise buildings, universities and industrial estates in Jakarta.

FIGURE 3 The Jakarta metropolitan area had a population of over 10 million in 2018 and a population density of over 14 000 people per square kilometre. It is the second largest urban area in the world.

8.5.3 Consequences of urbanisation

Growth of Jakarta

One of the consequences of urbanisation in Indonesia has been the dramatic growth of Jakarta, Indonesia's capital and largest city, located on the north-west coast of Java. The central island of Java is the world's most populous island, having a population density of 1000 people per square kilometre. The Jakarta Metropolitan Area (JMA) is now one of the world's largest urban areas. In 1930, Jakarta's population was around half a million people. By 1961 it had grown almost six-fold to 2.97 million. By 2005, it was almost 9 million. In 2019, the Special Capital Region of Jakarta had a population of almost 10 million, while the greater metropolitan agglomeration had a population of over 31 million.

FIGURE 4 Jakarta's urban growth

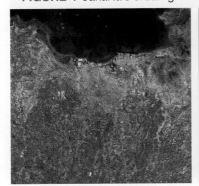
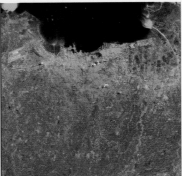

1976 (population 6 million)   1989 (population 9 million)   2014 (population 10.3 million)

 Resources

 Google Earth Jakarta, Indonesia

Loss of land

As Jakarta has become more urbanised, there has been a decrease in the amount of open green space — from nearly 30 per cent of the city's total area in 1984 to less than 10 per cent in 2015.

Prime agricultural areas have been lost and become residential and industrial areas. Urban land is worth more than agricultural land.

Environment

Indonesia's level of sewerage and sanitation coverage is very low. Sewage from houses and from industry, as well as industrial effluents and agricultural run-off, are polluting surface and groundwater. Air pollution levels are high, with traffic and industrial fumes combining with smoke from fires set by farmers and plantation owners in rural areas clearing forest lands for agricultural use.

Food production

Because young people, especially young men, migrate to Indonesia's cities in search of better job opportunities, there are fewer people taking over their families' farms. This could lead to the possibility of a food crisis if food production levels are not increased.

FIGURE 5 Smog over Jakarta

FIGURE 6 Traffic congestion in Jakarta

Job opportunities

Labourers who lived in Java and did not own land used to have very few sources of income. Now, most landless rural families on Java have at least one person working outside the village in a factory or service job. Today, less than 20 per cent of households depend on agriculture for their livelihood.

Subsidence

Land has been subsiding because more groundwater is being extracted, and also because of the additional load that the ground has to bear due to an increased volume of construction. Subsidence causes cracking of buildings and roads, changes in the flow of rivers, canals and drains, and increased inland and coastal flooding. In some parts of Jakarta, land has subsided by 1–15 centimetres per year — in other areas, this has been up to 28 centimetres per year.

New urban areas

New towns and large-scale residential areas have been developed in and around Jakarta. However, heavy flows of commuter traffic have led to increased levels of traffic congestion between the scattered new towns and the cities.

Explore more with my**World**Atlas

Deepen your understanding of this topic with related case studies and questions.
● Investigating Australian Curriculum topics > Year 8: Changing nations > **Urbanisation in Indonesia**

8.5 EXERCISES

Geographical skills key: GS1 Remembering and understanding **GS2** Describing and explaining **GS3** Comparing and contrasting **GS4** Classifying, organising, constructing **GS5** Examining, analysing, interpreting **GS6** Evaluating, predicting, proposing

8.5 Exercise 1: Check your understanding

1. **GS2** Explain, in your own words, why Indonesia has become very urbanised.
2. **GS2** Explain how and why Jakarta has become a major city within Indonesia and also on a world *scale*.
3. **GS2** Why do you think people have moved from rural areas to urban areas within Indonesia?
4. **GS2** What is the *interconnection* between the increasing population in Indonesia and the subsidence of land?
5. **GS1** Study **FIGURE 2**. Identify and list the cities that have become megacities (greater than 10 million people).

8.5 Exercise 2: Apply your understanding

1. **GS5** Refer to section 8.5.1.
   (a) What is Indonesia's current population? If the area of Indonesia is 1 904 569 square kilometres, what is its approximate population density?
   (b) How does this compare to Australia's population density of 3.1 people per square kilometre?
   (c) Describe, using statistics, how Indonesia has become very urbanised in a relatively short time.
2. **GS5** What do you believe are the three main reasons that Indonesia has undergone such rapid urbanisation? Give reasons for your choices.
3. **GS6** Which of the consequences of urbanisation do you think may continue to have the biggest effects on the *environment* in the future? Why? How important are these considerations to you?
4. **GS3** How is the urbanisation of Indonesia similar to and different from the urbanisation of another country you have studied, such as Australia, China or the United States?
5. Study **FIGURE 2**. In which time period did Jakarta experience its fastest growth? (*Hint:* Look at the width of the colour circle bands.)

Try these questions in learnON for instant, corrective feedback. Go to www.jacplus.com.au.

8.6 SkillBuilder: Creating and reading compound bar graphs

What are compound bar graphs?

A compound bar graph is a bar or series of bars divided into sections to provide detail of a total figure. These bars can be drawn vertically or horizontally. Compound bar graphs allow us to see at a glance the various components that make up the total.

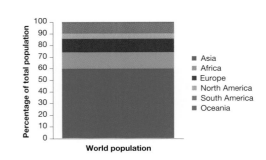

Select your learnON format to access:

- an overview of the skill and its application in Geography (Tell me)
- a video and a step-by-step process to explain the skill (Show me)
- an activity and interactivity for you to practise the skill (Let me do it)
- questions to consolidate your understanding of the skill.

on Resources

Video eLesson SkillBuilder: Creating and reading compound bar graphs (eles-1705)

Interactivity SkillBuilder: Creating and reading compound bar graphs (int-3285)

8.7 Characteristics of cities around the world

8.7.1 Urbanisation in South America

Megacities are primarily a phenomenon of the developing world, where their populations are increasing by between one and five per cent every year. If this continues, it is predicted that 40 megacities will exist by 2030 — and 21 of these will be located in the developing world, including countries in South America.

In recent years, the pace of urbanisation has been more rapid in South America than in North America and Europe. One hundred years ago, Buenos Aires was the only South American city with a population larger than one million. By 2015 there were 33 cities of this size. The five largest cities are São Paulo, Rio de Janeiro, Buenos Aires, Lima and Bogotá, with the first three defined as megacities.

These cities are a typical result of the urbanisation process occurring in South America, where the fastest population growth occurred between 1950 and the 1990s. The combined urban population of these five cities is nearly 73 million, one-fifth of South America's total urban population.

FIGURE 1 South America's urban population has kept up with total population growth.

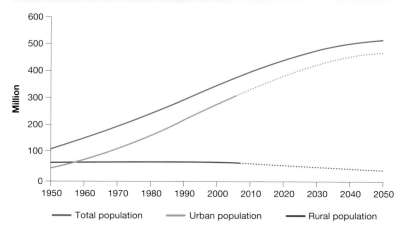

FIGURE 2 There are a number of very large cities in South America, which continue to increase in size.

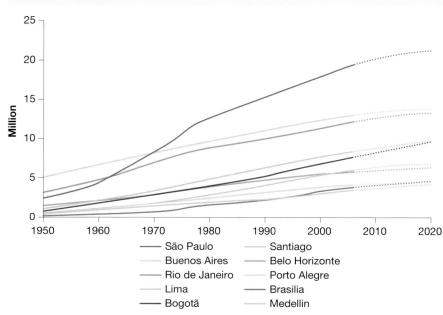

8.7.2 Urban issues in São Paulo, Brazil

São Paulo is located on a **plateau** on the top of an **escarpment** on the south-eastern coast of Brazil in South America (see **FIGURE 3**). From 1950 to 1980, São Paulo's population quadrupled from two million to more than eight million people. Since the 1980s growth has slowed, but it is still the largest city in South America. Its population is sprawled over an area of 7951 square kilometres (the city centre alone takes up an area of 1502 square kilometres). Compare this with Melbourne, which covers a total of 2453 square kilometres, and Sydney, with an area of 2037 square kilometres.

FIGURE 3 São Paulo sprawls into the distance; a sea of tall buildings.

The population density of São Paulo is 2469 people per square kilometre in the **metropolitan region** and 6832 in the inner-city district. São Paulo's population was 20 831 000 in 2014.

An average of 27 per cent of people in South America live in favelas (a term commonly used in Brazil meaning 'slums') — 28 per cent of the population in Brazil, 24 per cent in Argentina and over 36 per cent in Peru. This is a real challenge for these countries as they try to provide adequate housing, sanitation and other services to the urban poor.

São Paulo has become a major coffee producer, attracting workers and investors from throughout Brazil and many other countries. Today, many of the city's residents are direct or indirect descendants of immigrant groups including Italian, Portuguese, African, German, Lebanese and Japanese. São Paulo is home to the largest number of Japanese people outside Japan, the largest Lebanese population outside Lebanon, and the third largest Italian community outside Italy (after Buenos Aires and New York City).

TABLE 1 Growth of São Paulo's population from 1950 to 2030 (predicted)

Year	Population	Percentage of Brazil's urban population
1950	2 528 000	12.8
1955	3 521 000	13.7
1960	4 876 000	14.7
1965	6 380 000	14.8
1970	8 308 000	15.3
1975	10 333 000	15.5
1980	12 693 000	15.6
1985	13 844 000	14.4
1990	15 100 000	13.7
1995	16 469 000	13.2
2000	17 962 000	13.0
2005	19 591 000	12.9
2016	21 000 000	12.3
2030 (predicted)	23 444 000	12.0

Urban problems

São Paulo has grown rapidly and in an unplanned manner, leaving little space for highways and parks. Six million cars contribute to crippling traffic congestion and choking levels of air pollution in the city. South America has one of the highest car densities in the world. São Paulo is known for its chaotic traffic and in 2014 set a new record with a traffic jam stretching more than 344 kilometres during one peak hour. Some residents in outer city areas in São Paulo can spend between two and three hours each way commuting to and from work.

FIGURE 4 The built-up area can be clearly seen in this satellite image of São Paulo.

Air pollution levels in São Paulo are twice as high as those of New York City and London, even though Paulistanos (the name for people who live in São Paulo) have relatively low carbon emissions per capita.

8.7.3 Highways in the sky

Extreme wealth, as well as extreme poverty, exists in São Paulo. A number of wealthy elite live in luxury and avoid traffic congestion by travelling to and from work in helicopters. The rate of helicopter ownership in São Paulo is the highest in the world with around 700 registered helicopters taking up to 1300 flights per day in the city. This number is expected to continue to rise in the future.

FIGURE 5 The location of São Paulo in Brazil

Source: Spatial Vision

Living in poverty

Brazilian and overseas migrants who move to São Paulo with hopes of a better life often find it very difficult to find work and end up living in poverty. Around 3 million people live in favelas in São Paulo and surrounding areas. These favelas are located near gullies, on floodplains, on riverbanks, along railways, beside main roads and next to industrial areas.

Floods are common in São Paulo because there are very few green spaces to soak up the water. Air pollution is high and the two major rivers crossing the city are severely polluted, although these rivers are currently being cleaned up. The shortage and condition of the water supply are serious problems, especially for the urban poor living in favelas in São Paulo.

FIGURE 6 A topographic map of São Paulo

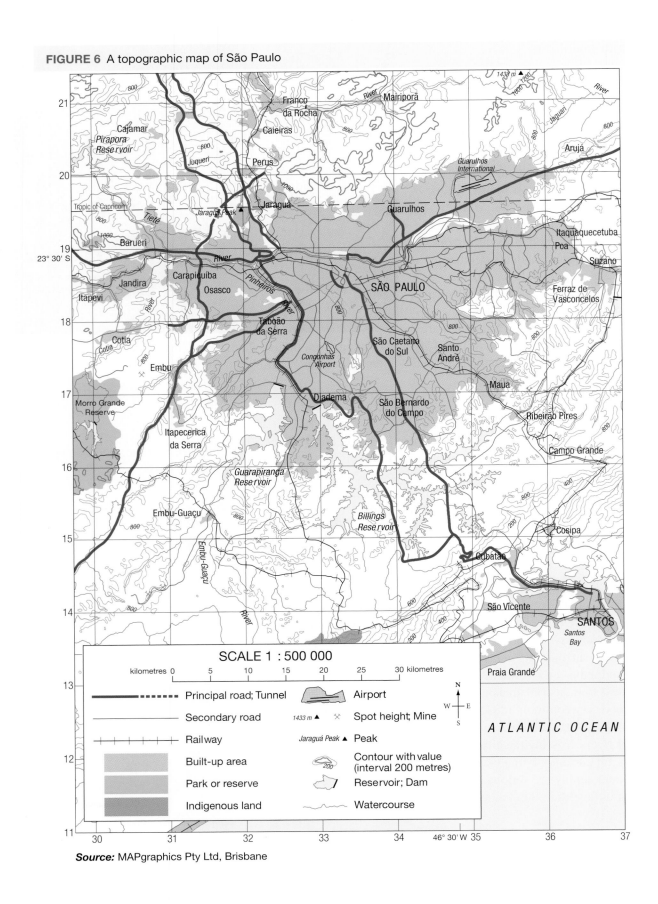

SCALE 1 : 500 000

kilometres 0 5 10 15 20 25 30 kilometres

Principal road; Tunnel
Secondary road
Railway
Built-up area
Park or reserve
Indigenous land

Airport
1433 m ▲ ✕ Spot height; Mine
Jaraguá Peak ▲ Peak
Contour with value (interval 200 metres)
Reservoir; Dam
Watercourse

Source: MAPgraphics Pty Ltd, Brisbane

FIGURE 7 Extremes of wealth and poverty in São Paulo. The Paraisópolis favela, home to between 80 000 and 100 000 people, is situated next to the gated complexes of the wealthy Morumbi district.

8.7.4 Cities in the United States

The distribution of major cities across the United States, including the largest cities (by population), is shown in **FIGURE 8**. The largest is New York City, New York, which is home to 8.6 million people. The second-largest city is Los Angeles, California, with a population of over 4 million; and the third-largest is Chicago, Illinois, with nearly 2.7 million people.

FIGURE 8 The distribution of major cities in the United States, 2018 (over one million people)

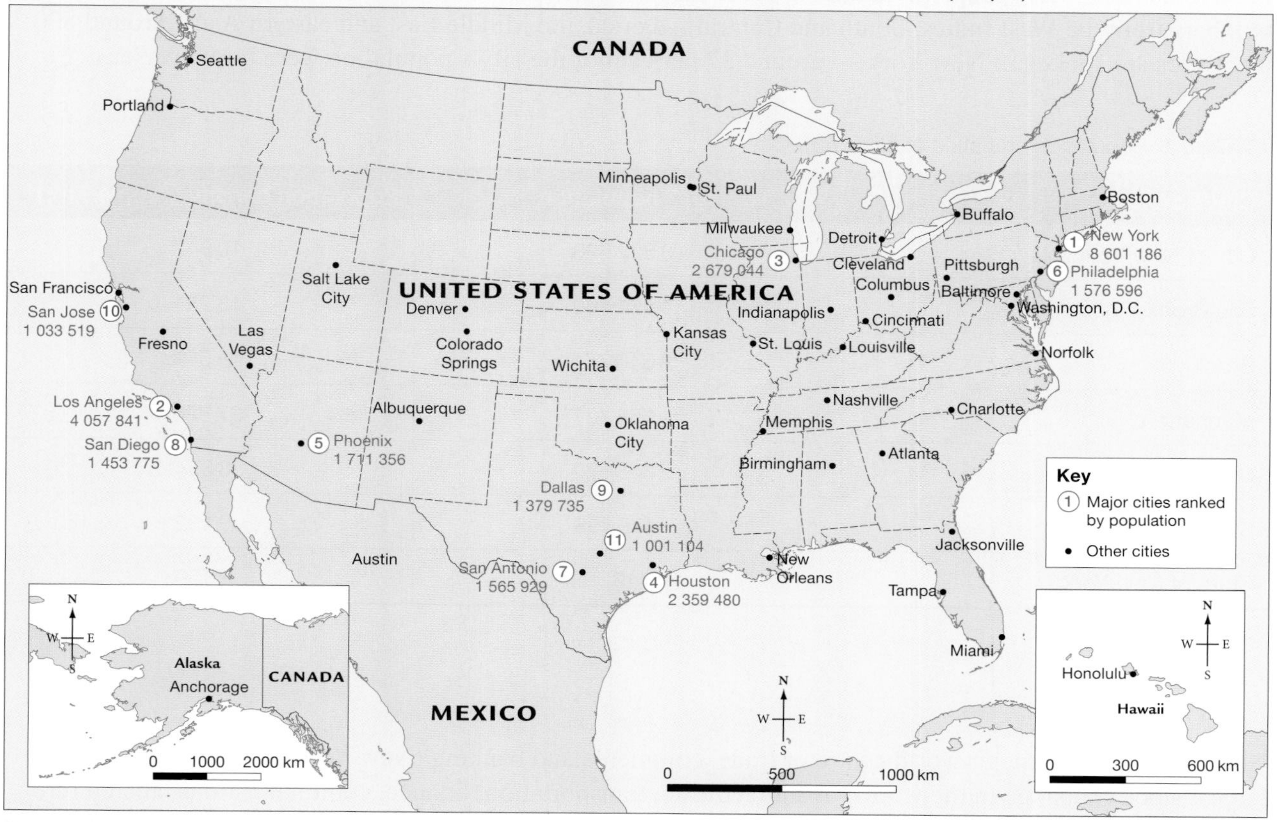

Source: Spatial Vision

8.7.5 New York City

The United States has a number of major cities distributed across the country. The largest of these is New York, one of the world's megacities, as its metropolitan area includes New York–New Jersey–White Plains. Its population in 2017 was 20.3 million.

In 1950 there were only two megacities, and New York was one of them (Tokyo in Japan was the other). In 2015, New York was the ninth-largest city in the world. By 2030 it is expected to be the thirtieth. There are only 11 states in the United States that are home to more people than New York City.

FIGURE 9 Manhattan Island and Central Park in New York

New York City is located on the eastern Atlantic Ocean at the mouth of the Hudson River. It is made up of five counties, or boroughs, separated by waterways — these are The Bronx, Brooklyn, Manhattan, Queens and Staten Island. Being located on four islands makes land very scarce and population density very high, at 11 084 people per square kilometre.

People

For many years, almost all immigrants came to the United States through New York City — and many of them remained. Many people living in New York are originally from European countries, but there are large numbers from the West Indies, South and Central America, the Middle East and eastern Asia. Around 800 languages are spoken in New York — around 37 per cent of the city's population were born overseas.

TABLE 2 Population statistics of New York City

City/Borough	2017 population	Population density (people/km^2)
City of New York	8 622 698	10 947
The Bronx	1 471 160	13 231
Brooklyn	2 648 771	14 649
Manhattan	1 664 727	27 826
Queens	2 358 582	8 354
Staten Island	479 458	3 132
State of New York	19 849 399	159

Economy

New York City is a major world centre of trade, commerce and banking (New York is also home to the largest stock exchange in the world), manufacturing, transportation, finance, communications, and culture and theatrical production. It is also the headquarters of the United Nations and a leading seaport.

FIGURE 10 Geographical characteristics of New York City

Source: Created from data from City of New York, New Jersey Department of Environmental Protection, New Jersey Geographic Information Network 2012

Boroughs

The Bronx County is the only part of New York that is connected to the US mainland. Historically there were many Irish and Italian migrants; today they are mostly Russian and Hispanic.

Brooklyn (also known as Kings County) is where most New Yorkers live; but Manhattan is the most densely populated county. It contains the highest number of skyscrapers, and includes Central Park and the village of Harlem. Central Park is nearly twice as large as the world's second-smallest country, Monaco.

┌─ Explore more with my**World**Atlas ─────────────────────────

Deepen your understanding of this topic with related case studies and questions.
- Investigating Australian Curriculum topics > Year 8: Changing nations > **Urbanisation in Australia and the USA**

8.7.6 What are European cities like?

European cities are old — many were first built by the Romans, and most existed during the Middle Ages. European cities are often smaller in scale and the buildings shorter than in the huge modern cities of North America and China. European cities are often described as *romantic, chic* or *picturesque*, words that would rarely be used to describe the cities of the United States or China.

Most European cities became cities 700–1000 years ago. They grew from being small-scale marketplaces, river crossings, road intersections, safe refuges and places of political power into the business, industrial and cultural centres they are today. Some of the largest and best known cities in Europe include London, Paris, Rome, Barcelona, Berlin, Milan, Vienna, Venice, Amsterdam and Prague.

A vibrant main square is a feature of European cities from Spain to Sweden and from England to Greece. The square was usually the site of a market place in medieval times, as well as being the communal and cultural centre of the city. Surrounding this square would be the most impressive buildings, such as the cathedral, town hall, concert hall, homes of the wealthiest families, museums and public monuments. It was the most prestigious place to live and to conduct business. The plaza, forum or market square was also an important meeting place for locals of all classes to mingle, gossip, find out local news and hold religious festivals.

As you can see in **FIGURE 11**, this square continues to bring pleasure to the local community and tourists alike. As a car-free space it is perfect for outdoor dining; and a weekly farmers' market is also held here. Interesting shops are located in the lower levels of the buildings and apartments, hotels and small offices on the upper floors.

Many cities in Australia and the United States have an area they refer to as their 'Little Italy' or 'Paris End'. These places usually have European-style features (see **FIGURE 12**) such as narrow laneways, outdoor dining, awnings, French or Italian restaurants and flower boxes.

FIGURE 11 This Italian piazza in Lucca occupies the site of a Roman amphitheatre. The curved row of buildings was built where the spectator stands once stood.

FIGURE 12 A laneway of restaurants and bars in Brussels, Belgium

In European cities, the tallest building is often a church. Even though some of the buildings crammed within the protective defences of the medieval city walls seen in **FIGURE 13** are less than 50 years old, they have been constructed to look the same as those built many hundreds of years earlier. The compact nature of European cities encourages wise use of space and encourages residents to walk, cycle or use public transport.

In Barcelona, Spain, the spires of the as-yet unfinished Basilica la Sagrada Familia, a huge Roman Catholic church that has been under construction for more than 100 years, dominate the city skyline in an older part of the city. It is being built in a region where the **population density** is greater than 50 000 residents per square kilometre (the city's highest). In a North American or Asian city, achieving a population density this great would be possible only with the building of residential skyscrapers. However, in this neighbourhood of Barcelona, the buildings are only five or six storeys high. Barcelona does have some very tall buildings, but they are found on the outer edges of the city and not in the older city centre.

FIGURE 13 The medieval quarter of a small French city

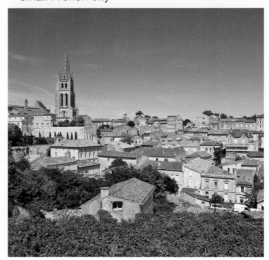

8.7.7 What is the future for European cities?

As they have developed, the ancient city centres of Europe have had to add water and sewerage systems and provide electric power, telephone and internet services as well as public transport access for their residents. The biggest issue in the past thirty years has been a huge increase in the level of car ownership. In parts of Paris, local people park their cars without applying the handbrake so that the vehicles can be pushed along by other drivers trying to fit their cars into very small parking spaces. The increased number of cars, even though many are small, has resulted in congestion and increased pollution.

To try to solve some transport problems, most European cities are trying to encourage people to walk, cycle and use public transport within the city. Many European cities, including London, Amsterdam, Paris, Barcelona and Copenhagen, have introduced public bicycle sharing schemes to provide people with an alternative to motorised transport, thereby helping to reduce traffic congestion, air pollution and noise.

FIGURE 14 A narrow French street that was not originally designed for car access or parking

FIGURE 15 Public share bikes in London

on Resources

Weblinks   Growth of São Paulo

BubbleUs

São Paulo

European traffic

Interactivity Urban USA (int-3120)

8.7 INQUIRY ACTIVITIES

1. Compare the area of São Paulo to that of the capital city in your state or territory. Trace an atlas map of your chosen city and add a *scale*. Now use the *scale* to draw an area of 90 square kilometres over the city area. This is the area of São Paulo. How does it compare to your city?   **Comparing and contrasting**
2. Study **FIGURE 4**. Describe the location of São Paulo and draw a sketch of the satellite image showing the area of the city. If Melbourne and Sydney were grid-shaped they would measure 40 and 33 square kilometres respectively. Now calculate the area that Melbourne or Sydney would take up and draw this over São Paulo. Compare the size of São Paulo with that of Melbourne and Sydney — write two statements to describe the differences.   **Comparing and contrasting**

3. Use the **Growth of São Paulo** weblink to learn how this city has grown over time. Where might future growth occur? Use the maps and images in this section to help you. **Evaluating, predicting, proposing**

4. Use the **BubbleUs** weblink in the Resources tab to brainstorm the urban problems experienced in São Paulo. Refer to the text and photographs in this section to help you get started. **Classifying, organising, constructing**

5. Study **FIGURE 6**.
   (a) Record the highest and lowest elevations in São Paulo's built-up area.
   (b) What evidence supports the fact that São Paulo is located on a plateau?
   (c) Describe where the steepest land is located.
   (d) Describe the rail and road routes from the coast to São Paulo. How can you explain the pattern shown?
   (e) Name one river that flows from the plateau to the sea and one that flows inland.
   (f) Use tracing paper to make a sketch of the built-up area of São Paulo, including the grid squares. Shade the built-up area and use the *scale* to calculate the area covered by one grid square on the map. Calculate the total area of São Paulo.
   (g) Describe the physical limitations to the growth of São Paulo. Predict where future urban growth will occur by shading areas on your map. Make notes on your map to justify why growth will occur in these locations and not in others. **Classifying, organising, constructing**

6. Use the **São Paulo** weblink in the Resources tab to watch the video.
   (a) Make a list of the projects that are trying to reduce urban problems in São Paulo. Rank these from 1 to 5, with 1 being the most effective and 5 the least. Justify your choices.
   (b) Compare your rankings with those of other students and discuss any similarities or differences. **Evaluating, predicting, proposing**

7. Conduct some research to find images of New York that reflect its characteristics. Use the information in this subtopic about its people and economy as well as the information in **FIGURE 10**. Include images of buildings, transport, culture and businesses, and produce a collage with labels. This might be in an electronic format or produced as a poster. **Classifying, organising, constructing**

8. Draw a sketch of **FIGURE 9**. Use the map to help you label Central Park and the Hudson River. In which direction is the photographer facing? **Classifying, organising, constructing**

9. What makes car ownership problematic for the residents of European cities? Use the **European traffic** weblink in the Resources tab, as well as evidence from the images in this subtopic, to support your answer. Use the concepts of *space*, *change*, *sustainability* and *scale* in your response. **Describing and explaining**

10. Investigate the city nearest to where you live to see whether it has a *place* influenced by European city design. *Hint:* Look for an area like Chinatown but European. How does the *environment* of this *place* reflect European cities? **Describing and explaining**

11. (a) Use your atlas and the internet to locate, on a base map of Europe, all the cities mentioned in this subtopic.
    (b) Annotate each city with its population size, the river that flows through it and one landmark found in that city.
    (c) Use a symbol to identify which cities have a public bicycle sharing scheme. **Classifying, organising, constructing**

12. How could European cities solve the problem of being overrun by cars? Produce a poster, brochure or PowerPoint presentation that fully explains the *change* required to implement your solution. **Evaluating, predicting, proposing**

8.7 EXERCISES

Geographical skills key: GS1 Remembering and understanding **GS2** Describing and explaining **GS3** Comparing and contrasting **GS4** Classifying, organising, constructing **GS5** Examining, analysing, interpreting **GS6** Evaluating, predicting, proposing

8.7 Exercise 1: Check your understanding

1. **GS1** Why have people been attracted to São Paulo?
2. **GS1** What is a favela? In which general areas are favelas located in São Paulo? Why do you think they are located in these *places*?

3. **GS2** Refer to **FIGURE 1** to describe South America's population growth. How does this compare to the population *change* in cities within the continent, shown in **FIGURE 2**?
4. **GS2** Study **FIGURE 5**. Describe São Paulo's location within both South America and Brazil.
5. **GS1** Name the five boroughs or counties that make up New York.
6. **GS1** In what year was New York one of the world's only two megacities?
7. **GS2** Describe the distribution of major cities in the United States. Where are most located?
8. **GS2** Describe New York's location within the United States and in terms of its natural geographical features. How have these features helped make New York a major city?
9. **GS1** List the terms used to describe a European market square.
10. **GS1** What features do most European cities have in common?
11. **GS2** How does the market square encourage *interconnections* between people and *places* in a European city?
12. **GS2** Explain why all the cities mentioned in this section were built near rivers.

8.7 Exercise 2: Apply your understanding

1. **GS4** Use **TABLE 1** to draw a graph showing the growth of São Paulo's population. When did the greatest growth take place? What percentage of Brazil's urban population lives there?
2. **GS6** Look at **FIGURES 3** and **7**. What do you think it would be like to live in such *environments*?
3. **GS4** Use the data in **TABLE 2** to draw a bar graph showing the population and densities of New York and its boroughs. Describe the pattern that you see.
4. **GS6** Use **FIGURES 9** and **10** and the graph you created in question 3 to write two sentences about population density in New York and where growth might occur in the future.
5. **GS5** Identify the *sustainable* and *unsustainable* features of European cities. Explain your answer.
6. **GS6** A European city such as Barcelona has regions of very high population density, even though the buildings are not as tall as those in a more recently developed city such as New York. How might the five-or six-storey buildings be able to contain so many living spaces?

Try these questions in learnON for instant, corrective feedback. Go to www.jacplus.com.au.

8.8 Creating sustainable cities

8.8.1 Sustainable urban solutions

Cities are huge consumers of goods and services. To be sustainable, cities need to develop so that they meet present needs and leave sufficient resources for future generations to meet their needs.

A sustainable city, or eco-city, is a city designed to reduce its environmental impact by minimising energy use, water use and waste production (including heat), and reducing air and water pollution.

Every city in the world experiences some type of problem that needs to be overcome — inadequate housing, urban sprawl, air and/or water pollution and waste disposal are just a few. Solutions to city problems have a better chance of succeeding if:
- responsibility is shared between governments, communities and citizens
- communities are involved in projects and decision making.

8.8.2 Sustainable urban projects

Urban greening program, Sri Lanka

Producing food in cities provides people with an income and improves local environments, as well as reducing the distance that food must travel to a consumer — '**food miles**'. With support from the Department of Agriculture, the Department of Education and the Youth Services Council, three city councils in Sri Lanka developed a program of community environmental management that led to the creation of 300 home gardens and 100 home-composting programs. It also helped organise and empower community groups, and the idea has now spread to many other municipalities in the country.

FIGURE 1 The urban greening program in Sri Lanka has been a success in many communities.

Beekeeping in urban areas

A worldwide movement of urban beekeeping has had beekeepers, in partnership with businesses and property owners in major cities, placing beehives on rooftops. The movement makes a strong connection between urban areas and food supply. This is happening in cities such as London (the Lancaster Hotel in London has its own hives, as does Buckingham Palace), New York, San Francisco, Paris, Berlin and Toronto. In Australia, there is a growing number of hives on city rooftops in Melbourne, Sydney and Brisbane.

FIGURE 2 Beehives on city rooftops, from where bees collect city pollen to make honey

Solar panels in Vatican City and Japan

Vatican City, Italy

Vatican City is the world's smallest independent state. In 2008 more than 2000 photovoltaic panels were fixed to the roof of one of the city-state's main buildings — the roof of the Paul VI Hall — enabling the Vatican to cut its carbon dioxide emissions by about 225 tonnes a year.

The 2400 panels heat, light and cool the hall and several surrounding buildings, producing 300 kilowatt hours (MWh) of clean energy a year. (see **FIGURE 3**).

Ota, Japan

Ota is located 80 kilometres north-west of Tokyo and is one of Japan's sunniest locations. Through investment by the local government, Ota is one of Japan's first solar cities — three-quarters of the town's homes are covered by solar panels that have been distributed free of charge.

FIGURE 3 Solar panels cover the roof of the Paul VI Hall, as seen from the dome of St Peter's Basilica.

FIGURE 4 A street in Ota, Japan — solar panels are visible on most of the houses.

Waste incineration in Vienna

A waste incineration and heat generation plant is part of a hard-waste management system in Vienna, Austria (see **FIGURE 5**). This plant became the first in the world to burn waste that cannot be recycled and use the energy generated by the plant in a heating network. The plant burns more cleanly and produces more heat and energy than many other waste generation plants, making it attractive to many urban communities. Each year, waste is turned into heat and electricity and supplies heating and hot water for 350 000 apartments — around a third of the city's total. The actual proportion of energy the waste supplies the city varies from season to season. Landfill waste has been reduced by 60 per cent in the city.

FIGURE 5 Spittelau waste treatment plant in Vienna, Austria. This power station burns waste, thus reducing landfill, to produce heat that is supplied to thousands of buildings.

The Loading Dock, Baltimore

The Loading Dock (TLD) is an organisation based in Baltimore, Maryland, in the United States, that recycles building material that was destined for landfill. The material is reused to help develop affordable housing while preserving the urban environment. The organisation works with non-profit housing groups, environmental organisations, local government, building contractors, manufacturers and distributors and uses human resources from within the community, improving living conditions for families, neighbourhoods and communities.

Since 1984, TLD has saved clients $40 million and diverted 12 000 truckloads of materials from the landfill.

Each year:

- TLD saves 421 kitchen cabinet sets, 68 km of timber and 1634 windows.
- TLD diverts 4489 doors from the landfill, which, if stood end-to-end, would be taller than 25 Empire State Buildings.
- The 83 000 carpet tiles saved from landfills would cover a football field.
- TLD could paint the exterior of the White House 85 times with the 98 000 litres of paint donated.
- TLD assists in the rehabilitation of nearly 15 000 homes. There has been interest in the project from 3000 other cities within the United States and in Mexico, the Caribbean, Hungary, Germany and five countries in Africa. All these projects will have a positive impact on people's lives and the urban environment.

DISCUSS

The projects described in this subtopic have been completed on a local *scale*. As a class, discuss why you think this might be the case. Do you think any of these *sustainable* city projects would work on a suburb-wide or city-wide *scale*? Why or why not?   **[Critical and Creative Thinking Capability]**

8.8 INQUIRY ACTIVITY

Work in pairs to identify one urban problem and design a *sustainable* program that would help to improve the condition. You will need to conduct some research to find similar problems and ways in which they have been tackled. Your program should include responses to the following.

- What is the urban problem? Include statistics (graphs or tables).
- Where is the problem located? Describe the location and include city/state/country map/s.
- What are the aims of your project? Describe what you hope to achieve.
- How will you achieve your aims? Describe your program or idea.
- Which individuals or groups are to be involved?
- What results would reflect success for your project?

Present your program to the class in the form of a Prezi or multimedia presentation, panel discussion or other format of your choice. Alternatively, you could share your programs through a class blog or wiki.

Evaluating, predicting, proposing

8.8 EXERCISES

Geographical skills key: GS1 Remembering and understanding **GS2** Describing and explaining **GS3** Comparing and contrasting **GS4** Classifying, organising, constructing **GS5** Examining, analysing, interpreting **GS6** Evaluating, predicting, proposing

8.8 Exercise 1: Check your understanding

1. **GS1** List the aims of a *sustainable* city.
2. **GS1** Which projects have the best chance of succeeding to overcome city problems?

8.9 Sustainable cities in Australia

8.9.1 Liveable Australian cities

Australian cities often perform well in worldwide rankings of liveability. In a 2018 survey, three Australian cities were ranked in the top 10 — Melbourne (2), Sydney (5) and Adelaide (10). Liveability is an assessment of the quality of life in a particular place — living in comfortable conditions in a pleasant location. But being liveable is not the same as being sustainable, which involves living in a way that sustains the environment and conserves resources into the future.

8.9.2 Measuring city sustainability

What makes a city sustainable? In 2010, the Australian Conservation Foundation conducted a study to measure the sustainability of Australia's 20 largest cities. The indicators measured were a combination of:

FIGURE 1 Darwin, the most sustainable city in Australia in 2010

- environment — air quality, ecological footprint, water, green building and biodiversity
- quality of life — health, transport, wellbeing, population density and employment
- resilience (the ability of a city to cope with future change): climate change, public participation, education, household repayments and food production.

The results showed that Darwin was the most sustainable city in Australia in 2010. It performed best in terms of the economic indicators of employment and household repayments. **FIGURE 2** shows the ranking for other cities.

FIGURE 2 A sustainability ranking of Australia's cities

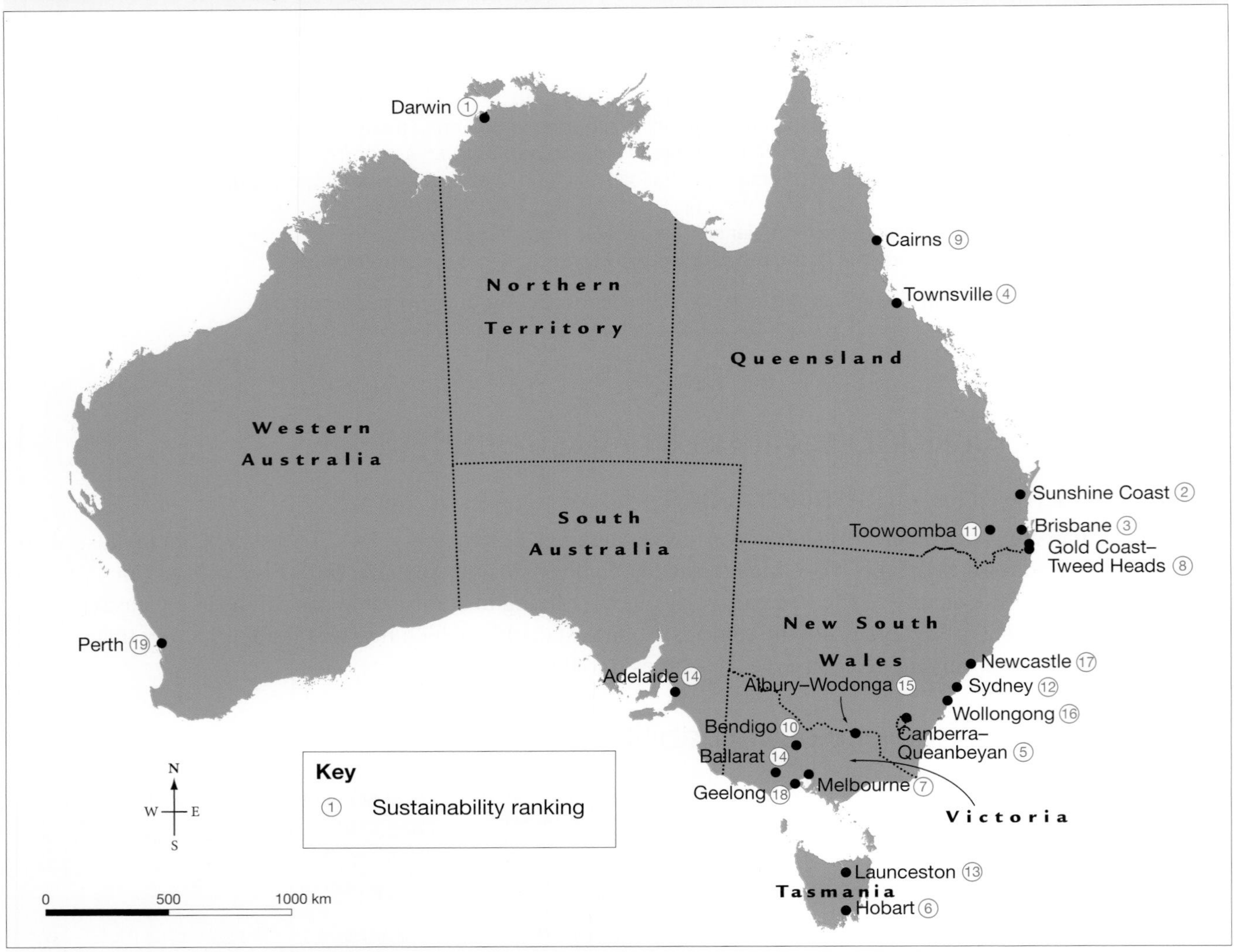

Darwin ①

Cairns ⑨

Townsville ④

Northern Territory

Queensland

Western Australia

Sunshine Coast ②

South Australia

Toowoomba ⑪
Brisbane ③
Gold Coast–
Tweed Heads ⑧

New South Wales

Perth ⑲

Adelaide ⑭
Albury–Wodonga ⑮
Newcastle ⑰
Sydney ⑫
Wollongong ⑯

Key
① Sustainability ranking

Bendigo ⑩
Ballarat ⑭
Geelong ⑱
Melbourne ⑦
Canberra–
Queanbeyan ⑤

Victoria

N
W — E
S

0   500   1000 km

Launceston ⑬
Tasmania
Hobart ⑥

Source: MAPgraphics Pty LTd, Brisbane

8.9.3 Local urban communities

In most cities, it is often action at a local community scale that can make the most difference in improving city sustainability. State governments and local councils have responsibility for improving complex infrastructure (for example, transport and water supply) for whole cities, but change at a local level can have positive results.

Sustainable communities in cities may have some of the following in common:
- friendly and social communities
- consume less energy and water and produce less waste
- have **medium-** to **high-density** rather than **low-density housing**
- are within walking distance of some public facilities and have excellent public transport links for longer trips
- include public places that people can walk to
- have good landscaping
- dwellings have been built to a budget to make them affordable.

FIGURE 3 The ACROS Fukuoka building located in Fukuoka, Japan

The ACROS Fukuoka building located in Fukuoka, Japan is an example of plants and greening being used to enhance a building (**FIGURE 3**). The terraced green roof and green walls merge with a park and contain around 35 000 plants. The green roof keeps the temperature inside more constant and comfortable, thus reducing energy consumption. It is also able to capture rainwater run-off and attracts many insects and birds. In addition, it is visually appealing and attracts many people to the surrounding park.

8.9.4 CASE STUDY: Christie Walk, Adelaide

Christie Walk is located in Adelaide in South Australia. It is a small urban village of 27 dwellings located on a quarter of an acre of land. The site is within easy walking distance of Adelaide's markets, parklands and CBD, which means car use is reduced. Around 40 people live at Christie Walk, ranging in age from very young to over 80 years.

A number of principles were used in the design of Christie Walk.
- Low energy demand (passive heating and cooling; natural lighting and sealed double glazing in all windows and glass doors)
- Maximising the use of renewable/solar-based energy sources (photovoltaic cells on the roof) and minimising the use of non-renewable energy sources
- Capturing and using storm water (in large underground rainwater tanks) and recycling waste water
- Creating healthy gardens and maximising the biodiversity of indigenous flora and fauna. The gardens also produce herbs, vegetables and fruit.
- Avoiding the use of products that damage human health
- Minimising the use of non-recyclable materials

FIGURE 4 One of the sustainable buildings in Christie Walk

FIGURE 5 A plan of Christie Walk in Adelaide

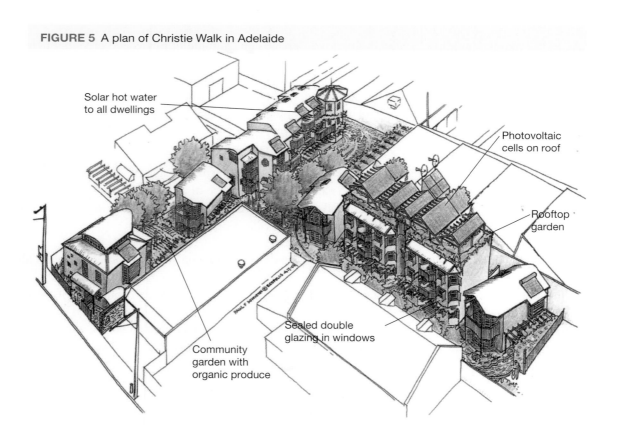

Solar hot water
to all dwellings

Photovoltaic
cells on roof

Rooftop
garden

Sealed double
glazing in windows

Community
garden with
organic produce

FIGURE 6 Rooftop gardens provide good insulation, protecting the buildings below from the hot sun in summer. In winter, they keep warmth from escaping from the building below.

 Resources

🔗 **Weblinks** Sustainable cities index

Sustainable Cities Awards

Explore more with myWorldAtlas

Deepen your understanding of this topic with related case studies and questions.
• Investigate additional topics > Urbanisation > **Brisbane: an eco-city**

8.9 INQUIRY ACTIVITIES

1. A hectare is equivalent to 10 000 square metres, or about 2.5 acres. In urban Australia, most houses were traditionally built on quarter-acre blocks (about 12 house blocks per hectare).
   (a) Walk around your neighbourhood or school area and pace out 100 × 100 metres. This gives you an idea of what one hectare looks like.
   (b) Use Google Earth or Google Maps to count or estimate the number of dwellings in your local area.
   (c) Compare your data with the definitions for low-, medium- and high-density housing. What type of housing density is in your local area? **Examining, analysing, interpreting**

▶

2. Use the **Sustainable cities index** weblink in the Resources tab to find out how cities in your state or territory have performed in measurements of sustainability. List five things that could improve sustainability in these cities. Is there anything you personally can do to make a difference?   **Evaluating, predicting, proposing**

3. Work in groups of three. Use the **Sustainable Cities Awards** weblink in the Resources tab to learn about projects that have won Sustainable Cities Awards in Australia. Each group member should read about three awards and summarise the projects to the others. Using a diamond ranking chart, rank the projects from most to least important for *sustainability*. Write the name or description of the best project in the top space of the diamond and the least *sustainable* at the bottom. Add the other projects to the chart after your group has discussed and agreed on the ranking.   **Classifying, organising, constructing**

4. Use ideas from this subtopic and further research to design a small *sustainable* urban neighbourhood. You may choose to work in groups or individually. You may like to use photographs of examples you find in your city/town or on the internet to draw your plan. Alternatively, video some examples and incorporate them into your design. Justify the inclusion of all the features you choose by annotating the plan or writing some notes to explain your choices. Present your final plan to the class as a panel presentation or on a class blog or wiki.   **Classifying, organising, constructing**

8.9 EXERCISES

Geographical skills key: GS1 Remembering and understanding **GS2** Describing and explaining **GS3** Comparing and contrasting **GS4** Classifying, organising, constructing **GS5** Examining, analysing, interpreting **GS6** Evaluating, predicting, proposing

8.9 Exercise 1: Check your understanding

1. **GS2** In your own words, describe the difference between a liveable and a *sustainable* city.
2. **GS1** List the indicators that were used to measure *sustainability* of cities in Australia.
3. **GS1** Why was Darwin voted the most *sustainable* city in 2010?
4. **GS1** Use **FIGURE 2** to find the direction and distance of Melbourne to Darwin.
5. **GS1** Study **FIGURE 2**. Which two cities are located the closest together?

8.9 Exercise 2: Apply your understanding

1. **GS1** Why is low-density housing considered *unsustainable* compared to medium- and high-density housing?
2. **GS2** Outline why a city's ability to produce food (with surrounding market gardens and farms) is a sign of resilience.
3. **GS4** Why is the ACROS Fukuoka (**FIGURE 3**) an example of a *sustainable* building?
4. **GS1** List four common characteristics of *sustainable* cities.
5. **GS2** Why is Christie Walk in Adelaide an example of a *sustainable* urban project?

Try these questions in learnON for instant, corrective feedback. Go to www.jacplus.com.au.

8.10 SkillBuilder: Constructing a basic sketch map

What is a basic sketch map?

A basic sketch map is a map drawn from an aerial photograph or developed during field work that identifies the main features of an area. Basic sketch maps are used to show the key elements of an area, so other more detailed characteristics are not shown.

Select your learnON format to access:

- an overview of the skill and its application in Geography (Tell me)
- a video and a step-by-step process to explain the skill (Show me)
- an activity and interactivity for you to practise the skill (Let me do it)
- questions to consolidate your understanding of the skill.

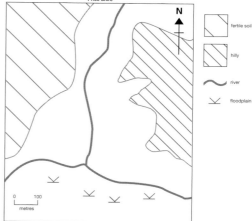

Resources

Video eLesson SkillBuilder: Constructing a basic sketch map (eles-1661)

Interactivity SkillBuilder: Constructing a basic sketch map (int-3157)

8.11 Thinking Big research project: One day in Jakarta, one day in New York City

SCENARIO

Your task is to plan an itinerary for someone visiting New York City and Jakarta for just one day each. What places can they visit that provide them with an experience of the life of these two cities? At the end of the day the visitor should have an understanding about the population characteristics, culture and environmental challenges for each city.

Select your learnON format to access:

- the full project scenario
- details of the project task
- resources to guide your project work
- an assessment rubric.

Resources

 projectsPLUS Thinking Big research project: One day in Jakarta, one day in New York City (pro-0174)

8.12 Review

8.12.1 Key knowledge summary
Use this dot point summary to review the content covered in this topic.

8.12.2 Reflection
Reflect on your learning using the activities and resources provided.

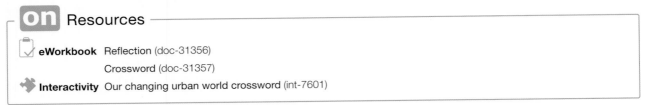

Resources

eWorkbook   Reflection (doc-31356)

Crossword (doc-31357)

Interactivity   Our changing urban world crossword (int-7601)

KEY TERMS

escarpment a steep slope or long cliff formed by erosion or vertical movement of the Earth's crust along a fault line

food miles the distance food is transported from the time it is produced until it reaches the consumer

high-density housing residential developments with more than 50 dwellings per hectare

low-density housing residential developments with around 12–15 dwellings per hectare, usually located in outer suburbs

medium-density housing residential developments with around 20–50 dwellings per hectare

megacity city with more than 10 million inhabitants

megaregion area where two or more megacities become connected as increasing numbers of towns and ghettos develop between them

metropolitan region an urban area that consists of the inner urban zone and the surrounding built-up area and outer commuter zones of a city

migration the movement of people (or animals) from one location to another

per capita income average income per person; calculated as a country's total income (earned by all people) divided by the number of people in the country

plateau an extensive area of flat land that is higher than the land around it. Plateaus are sometimes referred to as tablelands.

population density the number of people living within one square kilometre of land; it identifies the intensity of land use or how crowded a place is

pull factor favourable quality or attribute that attracts people to a particular location

push factor unfavourable quality or attribute of a person's current location that drives them to move elsewhere

sanitation facilities provided to remove waste such as sewage and household or business rubbish

slum a run-down area of a city characterised by poor housing and poverty

utilities services provided to a population, such as water, natural gas, electricity and communication facilities

9 Managing and planning Australia's urban future

9.1 Overview

Can Australia live in and grow its urban areas without making things worse for the future?

9.1.1 Introduction

We often hear the word *sustainable*, but what does it mean? Sustainability means meeting our own current needs while still ensuring that future generations can do the same. To make this happen, human and natural systems must work together without depleting our resources. Ultimately, sustainability is about improving the quality of life for all — socially, economically and environmentally — both now and in the future. In the words of HRH The Prince of Wales, 'Remember, our children and our grandchildren will ask not what our generation said, but what they did'.

Resources

| eWorkbook | Customisable worksheets for this topic |
| Video eLesson | Sustainable cities (eles-3495) |

LEARNING SEQUENCE

To access a pre-test and starter questions and receive immediate, **corrective feedback** and **sample responses** to every question, select your learnON format at www.jacplus.com.au.

9.2 Characteristics of sustainable cities

9.2.1 A common purpose

Our cities are facing an important challenge. Some predict that Australia's population will reach 45 million by 2050. If this is the case, then our cities must change and adapt to become more efficient in order to maintain or improve our current quality of life. How will we cope with a growing population?

Sustainable communities share a common purpose of building places where people enjoy good health and a high quality of life. A sustainable community can thrive without damaging the land, water, air, natural and cultural resources that support them, and ensures that future generations have the chance to do the same. The basic **infrastructure** should be designed to minimise consumption, waste, pollution and the production of greenhouse gases. Sustainable **urban** areas strike a delicate but achievable balance between the economic, environmental and social factors.

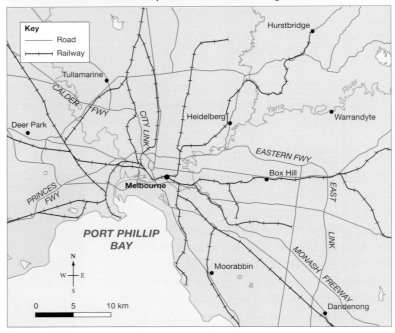

FIGURE 1 Melbourne, Victoria. Planning for sustainable use of Australian urban areas requires an understanding of scale.

Source: Department of Environment, Land, Water and Planning

A sustainable city is one that has a small **ecological footprint**. The ecological footprint of a city is the surface area required to supply a city with food and other resources and to absorb its wastes. At the same time, a sustainable city is improving its quality of life in health, housing, work opportunities and liveability.

We can address the challenges and opportunities for sustainable communities at two different scales: neighbourhood and city.

FIGURE 2 Street art in Melbourne, Australia

Ways to improve sustainability at the neighbourhood scale:

- reducing the ecological footprint
- protecting the natural environment
- increasing community wellbeing and pride in the local area
- changing behaviour patterns by providing better local options
- encouraging compact or dense living
- providing easy access to work, play and schools.

Ways to improve sustainability at the city scale:

- building strong central activities areas (either one major hub, or a number of specified activity areas)
- reducing traffic **congestion**
- protecting natural systems
- avoiding suburban sprawl and reducing inefficient land use
- distributing infrastructure and transport networks equally and efficiently to provide accessible, cheap transportation options
- promoting inclusive planning and urban design
- providing better access to healthy lifestyles (e.g. cycle and walking paths)
- improving air quality and waste management
- using stormwater more efficiently
- increasing access to parks and green spaces
- reducing car dependency and increasing walkability
- promoting green space and recreational areas
- demonstrating a high mix of uses (e.g. commercial, residential and recreational).

FIGURE 3 The Melbourne skyline with the Melbourne Sports and Entertainment Precinct in the foreground

Explore more with my**World**Atlas

Deepen your understanding of this topic with related case studies and questions.
- Investigate additional topics > Urbanisation > **Brisbane: an eco-city**

9.2 INQUIRY ACTIVITIES

1. How is an ecological footprint measured? Use the **Ecological footprint calculator** weblink in the Resources tab, or a teacher recommendation, to work through the steps to determine your own ecological footprint.

Examining, analysing, interpreting

2. After using the calculator, compare your ecological footprint with those of your classmates by creating a continuum on the board. It should start from smallest footprint (least planets consumed) to largest footprint (most planets consumed). Discuss which areas you think contributed to the wide variety of footprints.

Comparing and contrasting

3. Consider the areas listed in which a neighbourhood can become more *sustainable*. Create a table and, from your own perspective, detail the ways in which you believe your own suburb or neighbourhood is meeting these aims. Add another column and use the internet to research how your local council is trying to make your suburb more *sustainable*. Conclude by writing a few sentences to answer the following questions:
   (a) Is my neighbourhood *sustainable*?
   (b) How will liveability be improved?
   (c) What needs to change in order to make it even more *sustainable*?   **Evaluating, predicting, proposing**

9.2 EXERCISES

Geographical skills key: GS1 Remembering and understanding **GS2** Describing and explaining **GS3** Comparing and contrasting **GS4** Classifying, organising, constructing **GS5** Examining, analysing, interpreting **GS6** Evaluating, predicting, proposing

9.2 Exercise 1: Check your understanding

1. **GS1** Complete the following sentence: Some organisations have projected that Australia's population will reach _____ million by _____.
2. **GS1** What are the two main aims of a *sustainable* community?
3. **GS2** Explain the term *ecological footprint* in your own words.
4. **GS1** What are the two *scales* at which we can work to improve the *sustainability* of our communities? What are some of the differences between the two?
5. **GS2** What might a *sustainable* home look like to you?

9.2 Exercise 2: Apply your understanding

1. **GS2** Explain how the average Victorian's ecological footprint has *changed* in the last decade.
2. **GS2** Describe your understanding of what it means to have a good quality of life.
3. **GS6** Explain in a paragraph the simple actions everyone can make to reduce their ecological footprint.
4. **GS2** Explain how someone's quality of life can improve whilst still reducing his or her ecological footprint.
5. **GS2** Explain how increased green *space* and recreational areas can improve someone's quality of life.

Try these questions in learnON for instant, corrective feedback. Go to www.jacplus.com.au.

9.3 Sustainability of growing urban communities

9.3.1 The urban explosion

In 2008, for the first time in history, the world's urban population outnumbered its rural population. In 2019, the world's population exceeded 7.7 billion; it is expected to reach 9.2 billion by 2050. Where will all these people live? What challenges will cities and communities face in trying to ensure a decent standard of living for all of us?

Global population growth will be concentrated mainly in urban areas of developing countries. It is forecast that by 2030, 3.9 billion people will be living in cities of the developing world. The impact of expanding urban populations will vary from country to country and could prove a great challenge if a country is not able to produce or import sufficient food. Hunger and starvation may increase the risk of social unrest and conflict. On the other hand, farmers can help satisfy the food needs of expanding urban populations and provide an economic **livelihood** for people in the surrounding region.

One of the biggest challenges we face is ensuring that the sustainability of our economy, communities and environment is compatible with Australia's growing urban population (see **TABLE 1**).

TABLE 1 Percentage of population residing in urban areas by country, 1950–2050

	1950	1975	2000	2025	2050
Australia	77.0	85.9	87.2	90.9	92.9
Brazil	36.2	60.8	81.2	87.7	90.7
Cambodia	10.2	4.4	18.6	23.8	37.6
China	11.8	17.4	35.9	65.4	77.3
France	55.2	72.9	76.9	90.7	93.3
India	17.0	21.3	27.7	37.2	51.7
Indonesia	12.4	19.3	42.0	60.3	72.1
Japan	53.4	75.7	78.6	96.3	97.6
Papua New Guinea	1.7	11.9	13.2	15.1	26.3
United Kingdom	79.0	77.7	78.7	81.8	85.9

Source: UN Population Division 2011

9.3.2 The future for Australia

Australia's population will continue to grow and change. In particular, it will become more urban and its composition will age. Population increase threatens our fragile Australian environment. We continue to witness loss of biodiversity, limits on water supply, more greenhouse gas emissions and threats to food security. Our cities experience more traffic congestion and there are problems with housing availability and **affordability**. Access to services, infrastructure and green space are limited for some people in our communities. To handle these many challenges, we must plan effectively for an increased population by building communities that can accommodate future changes. This will build communities in which all Australians live and prosper.

9.3.3 The rural lifestyle

Approximately 93 per cent of Australia's growing population will be living in urban areas by 2050 (see **TABLE 1**). However, some urban residents will make a 'tree change' or a 'sea change' and relocate to rural areas or the coast. The population in rural communities is generally stable or decreasing, as many young people leave in search of jobs and study opportunities. Some rural communities manage to keep their populations stable by shifting their employment focus from manufacturing to services; by utilising better internet connections, to allow people to work remotely from their office; or by improving public transport links.

FIGURE 1 Percentage of world population living in urban areas, 2017

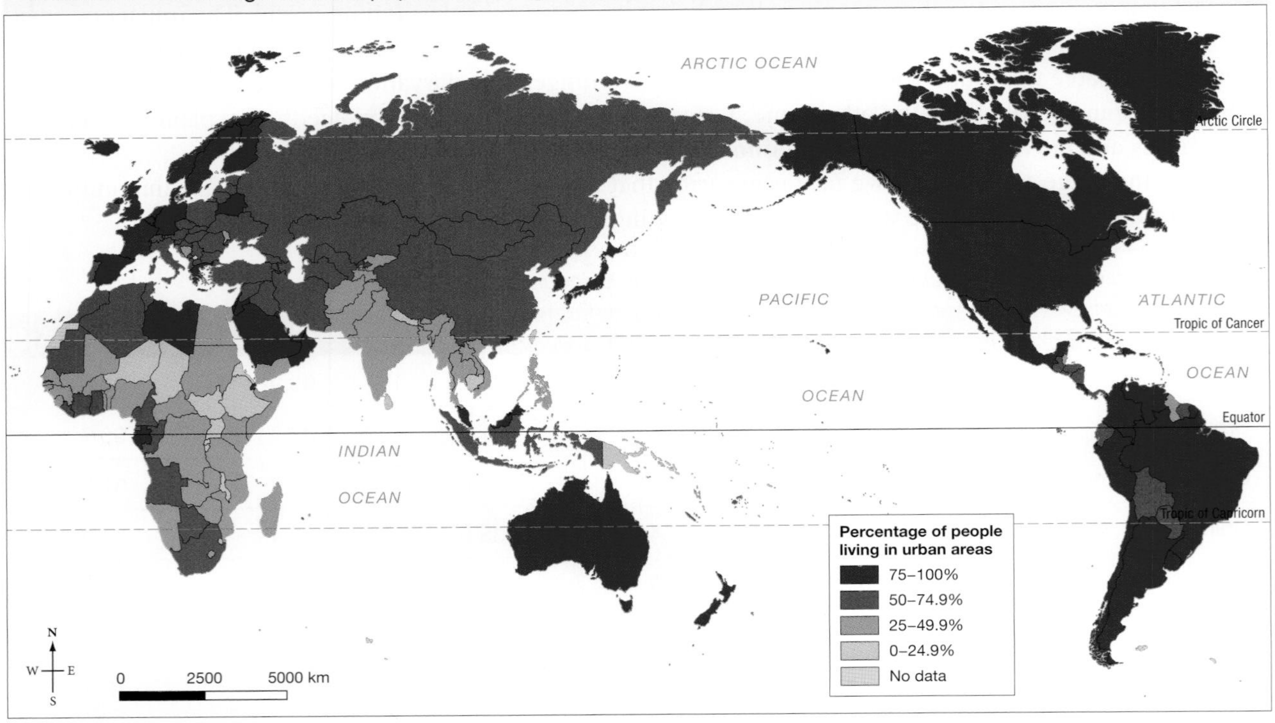

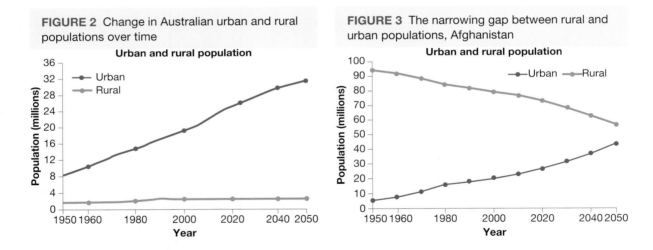

Percentage of people living in urban areas
- 75–100%
- 50–74.9%
- 25–49.9%
- 0–24.9%
- No data

Source: World Bank Data

FIGURE 2 Change in Australian urban and rural populations over time

Urban and rural population

- Urban
- Rural

Population (millions) vs Year (1950–2050)

FIGURE 3 The narrowing gap between rural and urban populations, Afghanistan

Urban and rural population

- Urban
- Rural

Population (millions) vs Year (1950–2050)

DISCUSS

Growing communities create growing problems. For example, social problems may include poverty, chronic unemployment, welfare dependence, drug and alcohol abuse, crime and homelessness. Working in small groups, brainstorm some of the impacts that growing communities may have on (a) the *environment* and (b) the economy.   **[Critical and Creative Thinking Capability]**

9.3 EXERCISES

Geographical skills key: GS1 Remembering and understanding **GS2** Describing and explaining **GS3** Comparing and contrasting **GS4** Classifying, organising, constructing **GS5** Examining, analysing, interpreting **GS6** Evaluating, predicting, proposing

9.3 Exercise 1: Check your understanding

1. **GS6** The global population is *changing*. Where do you think most of the world's population will live in the future?
2. **GS1** Is the pattern of population *change* the same or different in Australia?
3. **GS1** Where might the increased population live in Australia?
4. **GS2** Explain the social benefits of a sea *change* or tree *change*.
5. **GS5** Examine **FIGURES 2** and **3**. Explain the similarities and differences in the *changes* in urban and rural populations between the two *places*.

9.3 Exercise 2: Apply your understanding

1. **GS5** Refer to **TABLE 1**.
   (a) Which countries will be the most and least urbanised in 2050?
   (b) Which countries are predicted to experience the greatest percentage *change* in their urban population?
   (c) Are there any countries that have not seen a gradual increase in their percentage of urban population since 1950? Why might this be the case?
2. **GS4** Examine **TABLE 1**. Create a bar graph that shows the *change* over time for four countries of your choice.
3. **GS6** Young people leave rural areas in search of employment and education. What factors could contribute to you leaving the area where you live?
4. **GS6** In cities, we must face the challenges and opportunities of productivity, *sustainability* and liveability. If we address one goal, we can have an impact, either positively or negatively, on others. This demonstrates *interconnection*. For example, efficient public transport can fix congestion and improve access to jobs and opportunity (productivity). It can also reduce greenhouse gas emissions (*sustainability*) and make access to education, health and recreational facilities more affordable (liveability). Using the example of the National Broadband Network, how might productivity, *sustainability* and liveability be affected? Classify the effects you have listed as positive or negative.
5. **GS6** Evaluate the social, economic and *environmental* benefits and drawbacks of living in an urban area in Australia.

Try these questions in learnON for instant, corrective feedback. Go to www.jacplus.com.au.

9.4 SkillBuilder: Reading and describing basic choropleth maps

What is a choropleth map?

A choropleth map is a shaded or coloured map that shows the density or concentration of a particular aspect of an area. The key/legend shows the value of each shading or colouring. The darkest colours usually show the highest concentration, and the lightest colours usually show the lowest concentration.

Select your learnON format to access:

- an overview of the skill and its application in Geography (Tell me)
- a video and a step-by-step process to explain the skill (Show me)
- an activity and interactivity for you to practise the skill (Let me do it)
- questions to consolidate your understanding of the skill.

9.5 Managing our suburbs

9.5.1 Living on the edge

There is much at stake on the rural–urban fringe, with the conflict between farming and urban residential development reaching a critical point on the outskirts of Australia's cities. Australia is the driest inhabited continent on Earth, and just six per cent of its total area (45 million hectares) is arable land. The areas targeted by our state governments for residential development continue to expand. When some of our most fertile farmland is lost to **urban sprawl**, we reduce our productive capacity. Is this a recipe for sustainability?

On the edge of many Australian cities, new homes are being built as part of planned developments on greenfield sites. These were previously green wedges, wildlife habitats and productive farmland on the urban fringe. Accompanying these housing developments are plans for kindergartens, schools, parks, pools, cafés and shopping centres (often called amenities and facilities).

Having an 'affordable lifestyle' is the main attraction for people who purchase these brand new homes. They like the idea of joining a community and having the feeling of safety in their newly established neighbourhood.

Most new houses on the rural–urban fringe are bought by young first-home buyers, attracted by cheaper housing and greener surroundings. Generally, the residents of these fringe households feel that the benefits of their location outweigh the poor public transport provisions and long journeys to work and activities — trips that are usually made in a car.

9.5.2 Feeding our growing cities

Market gardens have traditionally provided much of a growing city's food needs, supplying produce to central fruit and vegetable markets. These 'urban farms' were located on fertile land within a city's boundaries but close to its edge, with a water source nearby and often on floodplains. They have been in existence in and around Australia's major cities since the 1800s, and some (such as Burnley Gardens in Richmond, Victoria) are now listed on the National Trust heritage garden register.

Fifty per cent of Victoria's fresh vegetable production still occurs in and around Melbourne, on farms such as those at Werribee and Bacchus Marsh. More than 60 per cent of Sydney's fresh produce is grown close to the city, with the bulk of it coming from commercial gardens such as those in Bilpin, Marsden Park and Liverpool.

These farms are important because:
- they provide us with nutritious food that does not have to be transported very far
- they provide local employment
- they preserve a mix of different land uses in and around our cities.

Currently, we can obtain our food from almost anywhere because we have modern transportation (such as trucks and planes), better storage technology (refrigeration and ripening techniques) and cheap sources (not necessarily the closest). However, this fails to recognise that Australia's population may double by 2050 and food will become more scarce on a global level. The eradication of our local food providers may be at our own peril.

Land use zoning is generally the responsibility of state planning departments but cooperation is required by all three levels of government: local, state and federal. We need to ensure that our green wedges are protected from becoming **development corridors**. The needs on both sides of the argument are valid. How can we house a growing population and provide enough food for them? Can we do both?

FIGURE 1 The history of Melbourne's urban sprawl

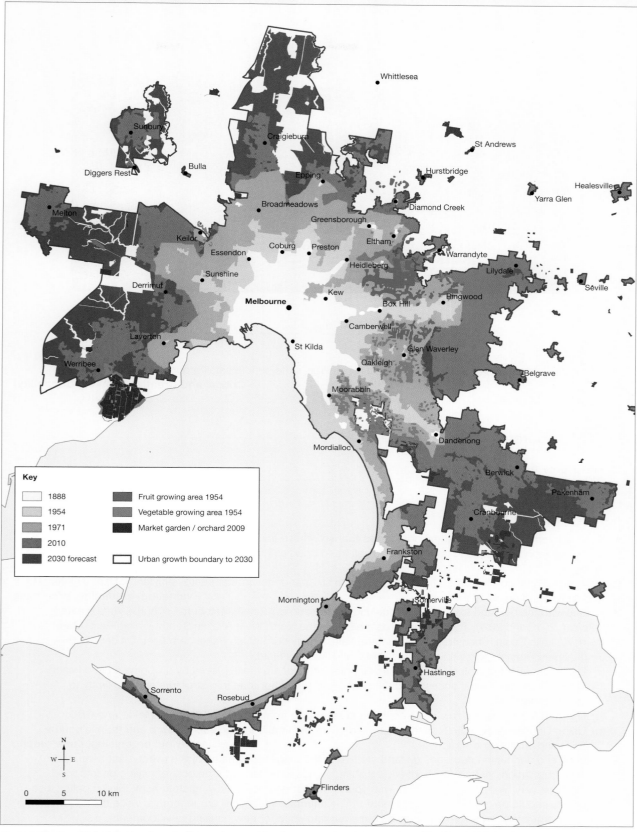

Key

1888	Fruit growing area 1954
1954	Vegetable growing area 1954
1971	Market garden / orchard 2009
2010	Urban growth boundary to 2030
2030 forecast	

Source: Various Victorian planning studies and current land use mapping. Map produced by Spatial Vision 2019.

FIGURE 2 The battle at our urban fringe: housing or farmland?

DISCUSS
'Sprawl is created by people escaping sprawl.' Discuss this statement in small groups.
[Critical and Creative Thinking Capability]

 Resources

⬧ **Interactivity** The 20-minute city (int-3122)

9.5 INQUIRY ACTIVITIES

1. Use the internet to research some companies that sell house and land packages in Victoria. What are some of the marketing messages that are used to sell the properties? Do you think they are able to deliver on their promises?   **Evaluating, predicting, proposing**

2. Many new homes on the urban fringe are built with six-star or seven-star energy efficiency. Use the internet to help you find out what this means.   **Evaluating, predicting, proposing**

3. To deal with the demand for affordable housing in Melbourne, 12 new Melbourne suburbs were created between 2019 and 2020. The new suburbs are Beveridge North West, Wallan South, Wallan East, Merrifield North/Kalkallo Basin, Shenstone Park, Lindum Vale, Craigieburn West, Pakenham East, Officer South Employment Precinct, Croskell, Kororoit Part 2 and Aviators Field.

    (a) Using a spatial technology tool such as Google Earth, locate these places by dropping a pin on each.

    (b) Assess these locations' access to public transport and major road infrastructure and proximity to essential services.

    (c) Using the ruler tool, calculate the approximate distance from each suburb to the Melbourne CBD.

    (d) Using Google Maps, calculate the estimated commute time between each suburb and the Melbourne CBD and Tullamarine Airport.   **Classifying, organising, constructing**

4. Housing and agriculture demands on land are two of the biggest dilemmas of the twenty-first century. A growing population needs to be housed, but it also needs to be fed, and the cost of relying on imported food can be very high. Set up a debate with your classmates on the following statement: 'Green belts close to the city should be preserved and protected.' The affirmative team will argue for this, while the negative team will argue that green belts should be removed and used for new housing developments.

                                   Evaluating, predicting, proposing

9.5 EXERCISES

Geographical skills key: GS1 Remembering and understanding **GS2** Describing and explaining **GS3** Comparing and contrasting **GS4** Classifying, organising, constructing **GS5** Examining, analysing, interpreting **GS6** Evaluating, predicting, proposing

9.5 Exercise 1: Check your understanding

1. **GS1** List the groups involved in the conflict over our rural–urban *spaces*.
2. **GS1** Why is it important for people to have rural *spaces*, such as market gardens, close to the city?
3. **GS1** Why is it important for cities to have access to more land for urban development?
4. **GS2** Refer to **FIGURE 1**. Describe how Melbourne's urban sprawl has *changed* in direction and pace.
5. **GS1** 'The eradication of our local food providers may be at our own peril.' Suggest what you think this statement means.

9.5 Exercise 2: Apply your understanding

1. **GS2** Explain the problems associated with urban sprawl with reference to locations around the city closest to where you live.
2. **GS2** Suggest some possible benefits of living in a newly created suburb.
3. **GS6** Predict how your own suburb or town might look as its population grows. List all the *changes* you may see due to increased population in your local area.
4. **GS6** Propose specific *changes* to transport infrastructure, which will prepare your own suburb or town for population growth.
5. **GS6** Predict how Melbourne's inner city will *change* with increased inner-urban population growth over the next decade.

Try these questions in learnON for instant, corrective feedback. Go to www.jacplus.com.au.

9.6 Managing traffic

9.6.1 The way forward

How did you get to school today? How long did you spend in the car? Were you stuck in a traffic jam? Australians who live in cities are experiencing longer commuting times than ever before, and this is only going to get worse. A growing population will mean an increase in cars — unless we start to tackle the problem from a sustainable perspective.

Road transport is a large source of greenhouse gas emissions in Australia (18 per cent), with passenger cars contributing more greenhouse gases than any other part of the transport sector. Some of the big issues in improving the sustainability of our transport systems are listed below.

Improving our infrastructure

Better public transport infrastructure will help improve the sustainability of our communities. Some cities have excellent rail systems or electrified tramways that were installed many years ago. But as cities grow and change, costly extensions may be required. Buses are much cheaper and quicker to upgrade. In Curitiba, Brazil, **bi-articulated buses** travel in dedicated bus lanes, and 70 per cent of the population uses the service. Public transport systems are cost-effective because it costs the same to run a bus or train with one passenger as it does with 1000 passengers. The more people who travel, the less it costs to transport each person.

Technologically advanced transportation

Since the late twentieth century, there have been many improvements in car design, occupant safety and fuel efficiency. In 2018, China sold over 1 million plug-in electric cars, and cities such as Paris, London and Melbourne are installing hundreds of car charging stations. Tesla motors has the largest share of the world's electric car market (50 per cent) but major car companies such as Volkswagen and Mercedes–Benz are investing billions of dollars and are therefore catching up. In Adelaide, the Tindo bus is powered by solar energy and can run for 200 kilometres between charges. More than 80 per cent of Brisbane's bus fleet runs on compressed natural gas (CNG) or meets the Euro V of VI diesel emissions standard.

FIGURE 1 How Australia compares on transport emissions

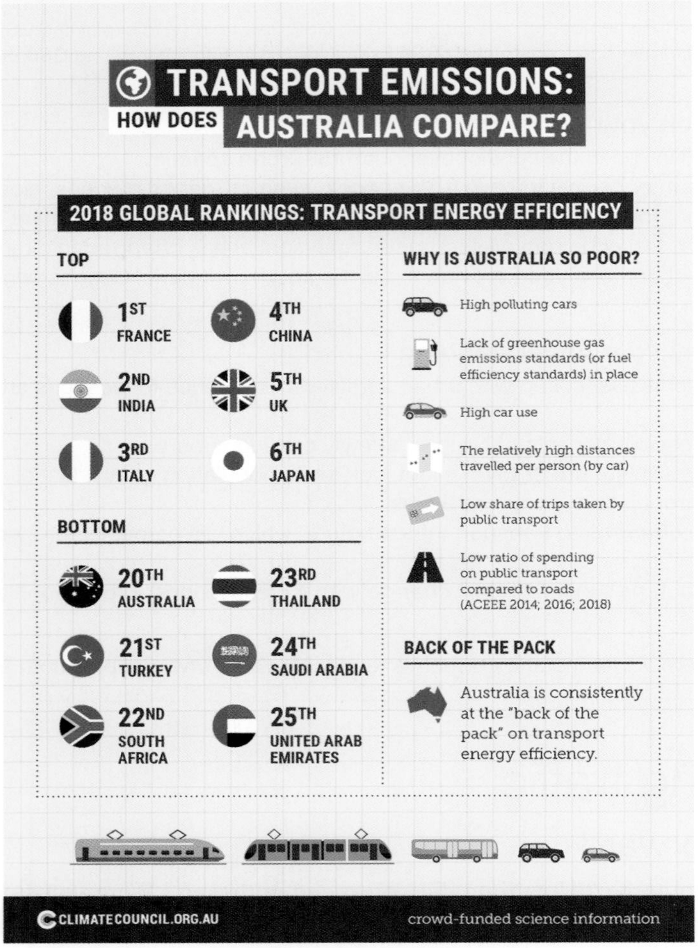

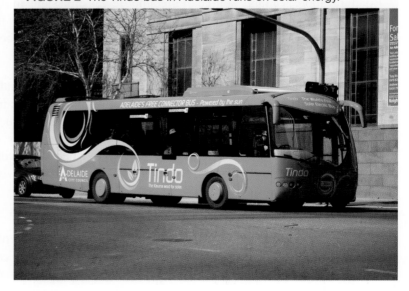

TRANSPORT EMISSIONS:
HOW DOES AUSTRALIA COMPARE?
2018 GLOBAL RANKINGS: TRANSPORT ENERGY EFFICIENCY
TOP
1ST FRANCE
2ND INDIA
3RD ITALY
4TH CHINA
5TH UK
6TH JAPAN
BOTTOM
20TH AUSTRALIA
21ST TURKEY
22ND SOUTH AFRICA
23RD THAILAND
24TH SAUDI ARABIA
25TH UNITED ARAB EMIRATES
WHY IS AUSTRALIA SO POOR?
High polluting cars
Lack of greenhouse gas emissions standards (or fuel efficiency standards) in place
High car use
The relatively high distances travelled per person (by car)
Low share of trips taken by public transport
Low ratio of spending on public transport compared to roads (ACEEE 2014; 2016; 2018)
BACK OF THE PACK
Australia is consistently at the "back of the pack" on transport energy efficiency.
CLIMATECOUNCIL.ORG.AU
crowd-funded science information

Denser urban settlements

FIGURE 2 The Tindo bus in Adelaide runs on solar energy.

ADELAIDE'S FREE CONNECTOR BUS - Powered by the sun
Tindo
ADELAIDE CITY COUNCIL

When an urban area is dense, the buildings are more compact, and more people live there. Dense urban settlements have 'efficiencies' already built in. Older cities, such as those in Europe, were established long before the invention of motor vehicles, meaning that they were built for walking. The older parts of European cities have narrow streets and laneways, and cannot cope with congestion. Europeans are less likely to own cars because they live close to their daily destinations, and this reduces the need for cars. In New York City, approximately 70 per cent of people travel to work by public transport, bicycle or foot and only about 50 per cent of families own a car. This is very different from the American average of 8 per cent of people who travel to work by public transport.

9.6.2 Changing our behaviour

Did you use a sustainable form of transport to get to school today? Cycling and walking are forms of mass urban transportation. Providing safe bike paths and walking routes makes people more likely to change their behaviours. If you have to travel by car, one way of increasing the effectiveness of each trip is car pooling. Governments or workplaces may also provide **incentives** for individuals to make a more sustainable transport choice.

FIGURE 3 Primary students catch the walking school bus.

Positive changes are happening, even if it is a little slow. The most recent figures show that approximately 10 per cent of Melbourne commuters either rode their bicycles to work or travelled on foot. This compares with approximately 4 per cent in 2001.

The toll we pay

Travel, particularly in our own cars, has increased at a rapid rate over the past 50 years. We have increased our mobility, independence and opportunities, and this has transformed the way in which land is used and people live. But as well as these benefits, car travel has created many health problems. Accidents and injury, climate change, air, water, soil and noise pollution, reduction in social interaction, and declining physical activity are all negative effects of car travel that take their toll on our health.

 Resources

Interactivity Smog buster (int-3123)

9.6.3 Why don't we just build more roads?

In an ideal world, a sustainable transport system would have a fast, clean, reliable and regular train service with waiting times of no more than ten minutes, day or night. Trams and buses would link into the train network, bringing people to the main parts of the system. Trams and buses would have priority over other traffic and run on the weekends. Station staff would be present at all times and the services would be safe and clean. What are some of the costs, other than financial, of using our cars instead of public transport?

Contrary to popular belief, building new roads and freeways does not actually ease congestion. This is because a new road simply becomes an opportunity for people to make new journeys that they may not have contemplated before; or they make the same journey more often; or they drive instead of taking public transport; or travel longer distances to accomplish the same task. All these things result in increased traffic on the new road, so the road system ends up just as congested as before. More energy and resources are consumed, and more pollution is generated.

FIGURE 4 Traffic jams slow down people and the economy.

FIGURE 5 This cyclist in China may be wearing a mask to reduce the effects of air pollution.

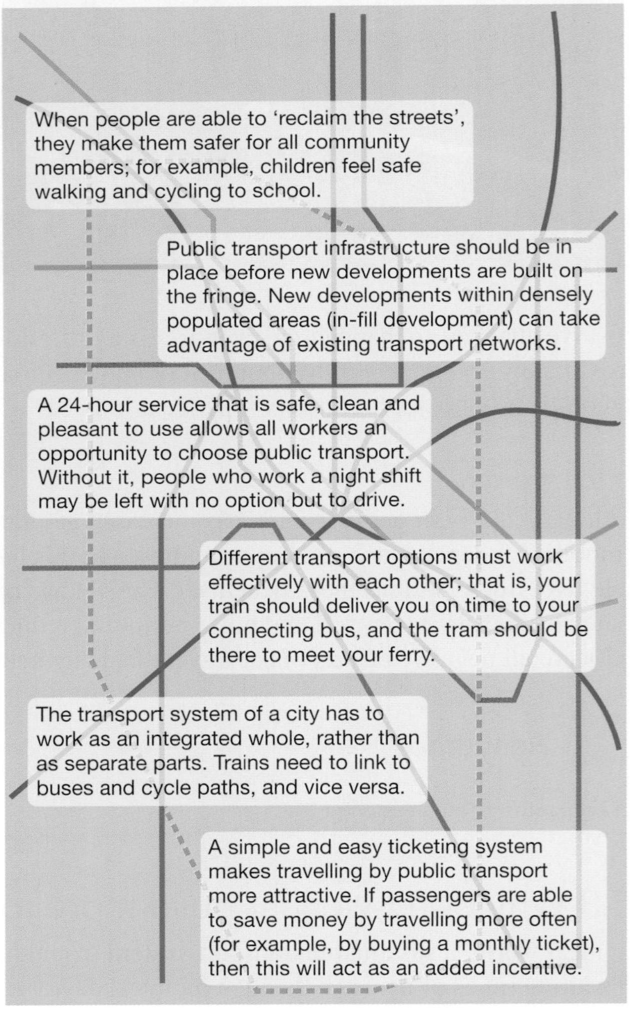

9.6.4 The benefits of an efficient public transport system

By shifting from car trips to public transport we can improve our **triple bottom line**. In other words, we improve economic efficiency, help the natural environment and do something good for society.

FIGURE 6 A model for transport in a sustainable city

When people are able to 'reclaim the streets', they make them safer for all community members; for example, children feel safe walking and cycling to school.

Public transport infrastructure should be in place before new developments are built on the fringe. New developments within densely populated areas (in-fill development) can take advantage of existing transport networks.

A 24-hour service that is safe, clean and pleasant to use allows all workers an opportunity to choose public transport. Without it, people who work a night shift may be left with no option but to drive.

Different transport options must work effectively with each other; that is, your train should deliver you on time to your connecting bus, and the tram should be there to meet your ferry.

The transport system of a city has to work as an integrated whole, rather than as separate parts. Trains need to link to buses and cycle paths, and vice versa.

A simple and easy ticketing system makes travelling by public transport more attractive. If passengers are able to save money by travelling more often (for example, by buying a monthly ticket), then this will act as an added incentive.

However, we also know that people will not get out of their cars and use public transport until public transport offers a high-quality, convenient and affordable service. Australia needs to make huge improvements in service frequency, connections and coverage. This formula has worked in other cities around the world and could work here in Australia.

Here in Australia we must look to develop Sustainaville — a community with its focus on public transport, walking and cycling.

 Resources

> ⬢ **Weblinks** Urban habitat
>
>   Crank busters
>
>   Transport urban myths

9.6 INQUIRY ACTIVITIES

1. What mode of transportation did you use to come to school today? How long did it take? How did your family members travel to their *place* of work or their school or university today? Use an internet mapping tool to help you work out how many kilometres your family travelled and by what means.
   Classifying, organising, constructing

2. Tally the results for your class's responses to question 1. Present the information in graph format. If possible, compare your results with another class. **Classifying, organising, constructing**

3. As a class, work out the minimum number of cars it would take to efficiently transport your entire class to school if everyone carpooled. **Classifying, organising, constructing**

4. Create a mind map of the way car travel affects your health, and then create a corresponding mind map of the way public transport affects your health. Include as many positive and negative points as you can with a brief explanation. **Classifying, organising, constructing**

5. Download a map of your suburb and print it out. Annotate it with current public transport options, such as trains, buses, bike paths and footpaths. Use different colours and a key to suggest improvements to existing options in your local *space*. **Classifying, organising, constructing**

6. There are many arguments for getting out of our cars and onto trams, trains, buses or bikes. Use the **Crank busters** and **Transport urban myths** weblinks in the Resources tab and other resources to prepare a class debate on one of the following topics.
   - People who own cars won't use public transport.
   - Bringing back tram conductors and station staff would increase fares.
   - Cars are more efficient than public transport.
   - Freeways reduce traffic congestion and pollution.

   You may be able to share the topics listed above among different groups and then present to the entire class. **Evaluating, predicting, proposing**

7. Curitiba in Brazil has installed a very successful bus rapid transit system (BRT), which has buses running about every 90 seconds and is used by 70 per cent of Curitiba's residents. Conduct some internet research using the **Urban habitat** weblink in the Resources tab or other sites, or view one of the many videos available online about the BRT system. Make a list of the unique features of the BRT and include some facts about the effect the system has had on the triple bottom line of Curitiba. How does this system compare to those you are aware of in your local community here in Australia? **Comparing and contrasting**

8. What kind of public transport system would you like to use? Design your own regional public transport option, using your local council area borders. Create a brochure showcasing the many benefits and features of the service. Include a map that details the routes of the service, frequency of service, hours of operation and cost. Use **FIGURE 6** to assist you. **Classifying, organising, constructing**

9.6 EXERCISES

Geographical skills key: GS1 Remembering and understanding **GS2** Describing and explaining **GS3** Comparing and contrasting **GS4** Classifying, organising, constructing **GS5** Examining, analysing, interpreting **GS6** Evaluating, predicting, proposing

9.6 Exercise 1: Check your understanding

1. **GS1** Which three areas does the triple bottom line concern?
2. **GS1** What are some of the negative effects of car travel?
3. **GS1** What is carpooling?
4. **GS2** Explain why public transport systems are cost effective.
5. **GS2** Explain the benefits of ride sharing for private transport.

9.6 Exercise 2: Apply your understanding

1. **GS3** Study **FIGURE 1**. How does Australia compare to the rest of the world in transport energy efficiency? What are some of the reasons for this?
2. **GS6** Consider the four areas for improvement listed in this subtopic. Which do you think will be the most important for (a) individuals and (b) the government to focus on in the next five years?
3. **GS2** The benefits of an efficient public transport system are many. If we were to discuss its impact on the **environment**, we would see less air and noise pollution, conservation of green **spaces** (public transport uses less **space** than roads), and reduced greenhouse gas emissions (GHGE). A full train produces about five times less GHGE than the cars needed to move the same number of people. Explain how an efficient public transport system would benefit the economy and society, following the **FIGURE 6** example to assist you.
4. **GS5** Study **FIGURE 6**. What does a public transport system need to be like in order to be a success?
5. **GS6** Propose methods to increase commuters of Melbourne's trains and trams.

Try these questions in learnON for instant, corrective feedback. Go to www.jacplus.com.au.

9.7 SkillBuilder: Drawing a line graph using Excel

What is a line graph?

A line graph is a clear method of displaying information so it can be easily understood. Using a digital means of drawing a line graph enables you to show multiple data sets clearly.

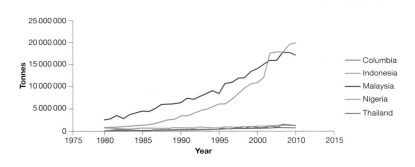

Go to your learnON title to access:

- an overview of the skill and its application in Geography (Tell me)
- a video and a step-by-step process to explain the skill (Show me)
- an activity and interactivity for you to practise the skill (Let me do it)
- questions to consolidate your understanding of the skill.

Resources

Video eLesson SkillBuilder: Drawing a line graph using Excel (eles-1662)

Interactivity SkillBuilder: Drawing a line graph using Excel (int-3158)

9.8 Sustainable cities

9.8.1 Masdar City

It may seem a little unusual to find a place like Masdar City in the Arabian Gulf. Masdar City, in the United Arab Emirates (UAE), was founded in 2006 to provide cutting-edge research into renewable and clean energy technologies. In this harsh and unforgiving climate, survival is all about sustainability, and resources must be used wisely in order to ensure a **viable** future.

The UAE possesses eight per cent of the world's oil reserves. By economic standards, it is a strong and stable country. The UAE government has recognised that, although it may have 100 years' worth of oil supplies left to sell to the rest of the world, it needs to ensure that, by the end of this century, its economy does not rely on its natural resources alone.

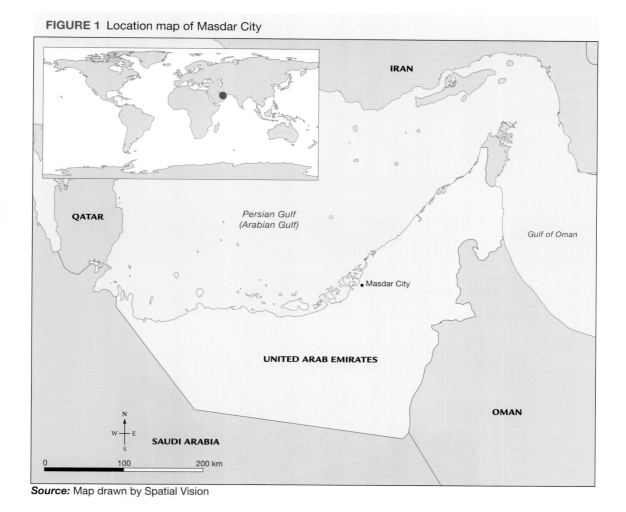

FIGURE 1 Location map of Masdar City

Source: Map drawn by Spatial Vision

The plan is for Masdar City to become a global leader in sustainability. Sustainable city-scale technologies and systems will be tested and then shared with other cities. This approach is intended to reduce the local and global ecological footprint of cities across the world.

It is intended for the city to have a population of 45 000 residents, and to make people, not cars, the focus. Pedestrians are king, streets are shaded by buildings or trees, and pleasant shaded walkways encourage walking. Masdar Plaza has 54, 30-metre-wide sunshades that open and close automatically at dawn and dusk. All these features aim to provide the highest quality working and living experience with the lowest possible environmental footprint.

FIGURE 2 Masdar City: a sustainable city in the desert

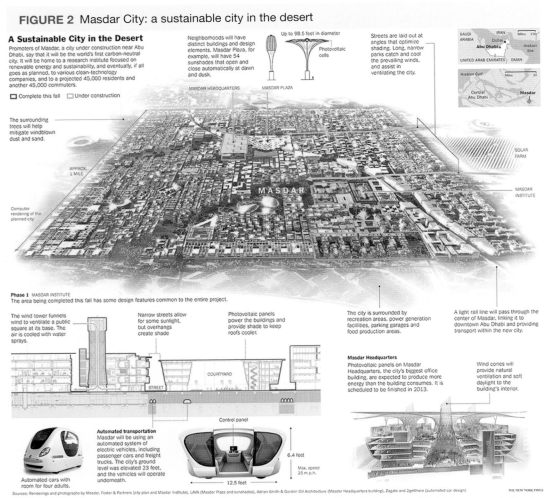

A Sustainable City in the Desert
Promoters of Masdar, a city under construction near Abu Dhabi, say that it will be the world's first carbon-neutral city. It will be home to a research institute focused on renewable energy and sustainability, and eventually, if all goes as planned, to various clean-technology companies, and to a projected 45,000 residents and another 45,000 commuters.

☐ Complete this fall ☐ Under construction

The surrounding trees will help mitigate windblown dust and sand.

APPROX. 1 MILE

Computer rendering of the planned city

Neighborhoods will have distinct buildings and design elements. Masdar Plaza, for example, will have 54 sunshades that open and close automatically at dawn and dusk.

Up to 98.5 feet in diameter

Photovoltaic cells

MASDAR HEADQUARTERS   MASDAR PLAZA

Streets are laid out at angles that optimize shading. Long, narrow parks catch and cool the prevailing winds, and assist in ventilating the city.

MASDAR

SOLAR FARM

MASDAR INSTITUTE

Phase 1 MASDAR INSTITUTE
The area being completed this fall has some design features common to the entire project.

The wind tower funnels wind to ventilate a public square at its base. The air is cooled with water sprays.

Narrow streets allow for some sunlight, but overhangs create shade

Photovoltaic panels power the buildings and provide shade to keep roofs cooler.

COURTYARD

STREET

The city is surrounded by recreation areas, power generation facilities, parking garages and food production areas.

A light rail line will pass through the center of Masdar, linking it to downtown Abu Dhabi and providing transport within the new city.

Masdar Headquarters
Photovoltaic panels on Masdar Headquarters, the city's biggest office building, are expected to produce more energy than the building consumes. It is scheduled to be finished in 2013.

Wind cones will provide natural ventilation and soft daylight to the building's interior.

Control panel

6.4 feet

Max. speed 25 m.p.h.

12.5 feet

Automated transportation
Masdar will be using an automated system of electric vehicles, including passenger cars and freight trucks. The city's ground level was elevated 23 feet, and the vehicles will operate underneath.

Automated cars with room for four adults.

Sources: Renderings and photographs by Masdar, Foster & Partners (city plan and Masdar Institute), LAVA (Masdar Plaza and sunshades), Adrian Smith & Gordon Gill Architecture (Masdar Headquarters building), Zagato and 2getthere (automated car design)

THE NEW YORK TIMES

FIGURE 3 A newly completed courtyard in Masdar City

9.8.2 Auroville

Auroville is a planned community in south-east India for up to 50 000 people, which has been under development since its inception in 1968. As of 2019, more than 2700 people from over 50 nations live and work in Auroville. It is located close to the Coromandel Coast, 10 kilometres north of Pondicherry and 150 kilometres south of Chennai (see **FIGURE 4**).

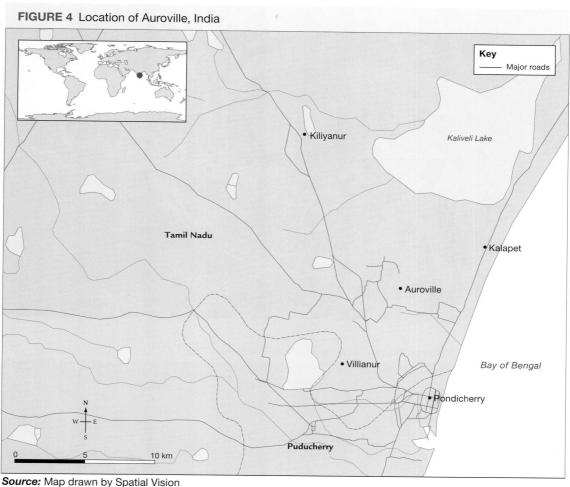

FIGURE 4 Location of Auroville, India

Source: Map drawn by Spatial Vision

The Auroville vision

Auroville wants to be a universal town where men and women of all countries are able to live in peace and progressive harmony above all creeds, all politics and all nationalities. The purpose of Auroville is to realise human unity.

Source: Mirra Alfassa, the 'Mother' of Auroville

The grand plan for Auroville was to create two geographical regions around the Matrimandir — a gleaming dome-shaped building that is the spiritual centre of Auroville (see **FIGURE 6**). The city area contains residential houses and community centres. The planned green belt is a forest that surrounds the city area. It has two functions: adding greenery and beauty, and to be a source of food and raw materials. Residential quarters within the city area are organised into self-contained communities with shared local water and wastewater systems (although the entire city shares an electricity supply).

FIGURE 5 A plan of Auroville, India

FIGURE 6 The Matrimandir is at the centre of the Auroville spiral plan.

Sustainable features of Auroville

Solar technology

- Water pumping and heating, street lighting and electricity generation all use solar power.
- A 1000-meal per day solar kitchen is powered by a solar concentrator. The design uses hundreds of mirrors to focus sunlight onto a heat receiver. The coils around the heat receiver are filled with water and, when the water turns to steam, it is used for cooking.

Water technology

- Waste water is treated at individual households and local communities, rather than at large sewage treatment plants.
- Rainwater harvesting is an important source of fresh water.

Revegetation

- The site chosen for Auroville was an eroded plateau that was suffering from desertification. Two million trees have been planted, and the area is now a green and forested landscape.

Community recycling and reuse projects

- Examples of recycling projects include the Auroville central exchange shop. Instead of dumping old and unwanted items in the rubbish, community members are encouraged to exchange or donate these items to the shop, since another person may find the item useful.
- Auroville has provided social and economic benefits for the surrounding villages. More than 5000 villagers are hired from nearby villages as cleaners, construction workers and maintenance workers, and are given job training. This has increased family incomes substantially and improved the standard of living within the communities.

Resources

📄 **Digital document** Masdar City infographic (doc-11470)
🔗 **Weblink** Auroville

9.8 INQUIRY ACTIVITIES

1. Locate Abu Dhabi and your home town or city on a world map. Describe the location of each *place*, including the latitude and longitude of each.   **Describing and explaining**

2. Design your own *sustainable* city, using the image of Masdar City in **FIGURE 2** as a guide. Ensure you provide:
   - a map of the city, noting important features
   - an inset map showing potential location (country and continent, with some reference to climate)
   - information on *scale*
   - information on how the city generates its own energy sources
   - a list of water efficiency measures
   - descriptions of green *spaces*
   - an outline of transport options provided to residents.

   You could choose to create a model in a small group or a blueprint on paper, using ICT to assist you.
   Classifying, organising, constructing

3. Conduct internet research to find out more about the solar kitchen at Auroville, or other examples around the world. Create a diagram that shows how heat is generated by the solar bowl concentrator, which cooks the meals in the shared kitchen.   **Describing and explaining**

4. Use the **Auroville** weblink in the Resources tab to find out more about Auroville.
   Examining, analysing, interpreting

9.8 Exercise 1: Check your understanding

1. **GS1** Refer to **FIGURE 1**. Which countries border the United Arab Emirates?
2. **GS6** In what activities might the residents of Masdar City be engaged?
3. **GS6** The Arabic word *masdar* means 'the source'. Why do you think the city was given this name?
4. **GS3** What type of climate exists in Abu Dhabi? How does it compare with the climate of where you live?
5. **GS1** What was the Auroville site like before 1968? How has the site *changed* over time?
6. **GS2** Explain the principles behind developing the community of Auroville. Is this how most cities or communities are planned? Why or why not?
7. **GS2** Would a solar kitchen be useful in a school setting? Justify your response.

9.8 Exercise 2: Apply your understanding

1. **GS6** What do you think it would be like to be a teenager living in Masdar? Would you like to live there? Why or why not?
2. **GS4** Study **FIGURE 2**, also available as the 'Masdar City infographic' resource (doc-11470) in the Resources tab. Create a table that shows the economic, social and *environmental* benefits of Masdar City.
3. **GS6** Masdar City was master-planned with many efficiencies built into the design. Do you think it's easier to design a city from scratch or to make changes to an existing city in order to make it *sustainable*? Explain your response.
4. **GS3** How is the development of Auroville different from that of Masdar? Are they both trying to achieve the same outcome? How are they each proposing to reach their goals? Use a Venn diagram to compare and contrast.
5. **GS6** Would a community like Auroville succeed here in Australia? Why or why not?

Try these questions in learnON for instant, corrective feedback. Go to www.jacplus.com.au.

9.9 Planning for a sustainable and liveable future

9.9.1 Higher-density living, smaller households

Australian cities are experiencing an apartment revolution. More people are choosing to live in the centre of cities in high-rise apartments rather than in houses on big suburban blocks. Urban life now sees families and individuals moving to the inner city for a variety of reasons, such as seeking to make a smaller ecological footprint, or avoiding long commutes to school, work and shops.

Australian households are changing in structure all the time, and recent data suggest the greatest increase will be in **family households**, which will grow from 6.5 million in 2016 to around 9.2 million in 2041 and will remain the most common type of household in Australia. Single-person households are projected to rise from 2.3 million in 2016 to 3.5 million in 2041. This is due to two main factors: the ageing Australian population, where women predominantly outlive men, and the fact that many adults are delaying marriage and starting a family much later.

9.9.2 Going green

All housing can be designed to be sustainable. However, medium- and higher-density housing can offer the greatest opportunity for energy savings. Buildings with shared walls and more than one storey (such as two-storey and semi-detached homes, terraces and apartments) use less energy for heating and cooling than single-storey detached homes.

In Australia, people have started to value being able to walk to facilities and workplaces, so our urban centres are increasing in population density. For business and residential purposes, urban sprawl is far less sustainable than high-rise buildings. A sustainable building may include on-site energy generation (such as solar panels and wind turbines) and passive energy design (such as insulation), reducing the need for air-conditioning and heating. 'Green' or recycled building materials can also lower the environmental costs of construction.

FIGURE 1 Council House 2 in Melbourne was the first Australian building to be awarded a six-star green rating. Its features include a green roof and louvred shading system.

Green roofs and walls

Green roofs and walls have a history dating back thousands of years. People are rediscovering the benefits of creating healthy, green buildings. A green, or living, roof is a roof surface that is planted partially or completely with vegetation over a waterproof layer. They may be extensive, with simple ground-cover vegetation, or intensive, with soil more than 200 millimetres deep and planted with trees. Green walls are external or internal walls of buildings that include vegetation, either in stacked pots or in growing mats.

FIGURE 2 The Burnley Living Roof at the University of Melbourne's Burnley Campus

Green roofs have several benefits. They:

- are aesthetically pleasing
- provide a cooling effect on local microclimate
- reduce carbon dioxide (CO_2)
- reduce air pollution
- provide insulation for buildings
- provide recreational space for local residents and workers
- allow for fresh food to be grown close to where it's needed, thus reducing transport costs.

The high life

In the last century, Europe has transformed itself from a largely rural to a mostly urban continent. It is estimated that in 2019, approximately 75 per cent of Europe's population (550 million people) lived in urban centres of more than 5000 people. About two-thirds of energy demand is linked to urban consumption and up to 70 per cent of CO_2 emissions are generated in cities. The urban way of life is both part of the problem and part of the solution. The density of urban areas allows for more energy-efficient forms of housing, transport and other services. Consequently, measures to address climate change may be more efficient and cost-effective in big, compact cities than in less densely built spaces.

9.9.3 Planning for a liveable future

Managing and planning Australia's future urban areas will take the efforts of many. We, as citizens of Australia and the world, must be prepared to make significant changes to the way we live if we wish to enjoy a good quality of life in the future. Sustainability and liveability must be on the agenda for governments, communities and individuals.

The role of governments

Governments can commit to sustainability in a number of ways. They may offer incentives such as **rebates** on solar panels or water-efficient showerheads. They can fund research into sustainable technologies. Governments can adopt strict planning regulations and well-defined urban growth boundaries. They can have clear policies on levels of air quality, business sustainability, and the construction or **retrofitting** for sustainability of 'green' buildings. They can develop land-use plans that encourage sustainability and biodiversity.

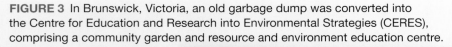

FIGURE 3 In Brunswick, Victoria, an old garbage dump was converted into the Centre for Education and Research into Environmental Strategies (CERES), comprising a community garden and resource and environment education centre.

The role of communities

Communities and organisations are working with governments, businesses and individuals to respond to global challenges such as climate change. There are many measures in place to improve transport and mobility, develop effective use of our land, and plan and develop appropriate policies.

Communities maintain and improve infrastructure and open spaces, and can help us work at the neighbourhood level to build a more sustainable community. The Sustainable Street Approach has seen the emergence of hundreds of Sustainable Villages around Australia. These villages are run by local residents who work together to improve local liveability. They might establish community gardens or purchase solar systems in bulk. Some great examples of communities working with governments to improve liveability and sustainability are shown in **FIGURES 3**, **4** and **5**.

FIGURE 4 Green roof design on apartment buildings in Sydney's Pyrmont.

The role of the individual

We can all seek to enjoy a quality of life that does not damage the environment. Although you might feel powerless, in the next decade you will be making your own contribution to society and thinking about what kind of world you would like to grow old in. You will need to consider your sustainable choices in the action areas shown below. What is *your* personal sustainability plan? Ultimately, if you want to improve your quality of life and the environment, make your choices sustainable ones. You could get involved by:

- riding or walking to school each day
- establishing an eco-classroom at your school
- learning more about the connections between Aboriginal and Torres Strait Islander peoples and their land
- installing solar hot water or solar panels at your residence
- growing your own food.

FIGURE 5 A vertical garden disguises a five-storey car park at Southbank, Victoria

FIGURE 6 Action areas

Energy

Aboriginal and Torres Strait Islander knowledge

Waste

Community

Transport

Sustainable purchasing

9.9 INQUIRY ACTIVITIES

1. Green roofs can be built anywhere. Select a rooftop on a building at your school, and create a plan for your own green roof. To find inspiration, conduct research on successful green roofs around the world. You will need to include a design, information on size and materials needed, and how and why it would be accessed. Present your design using a program such as Prezi or PowerPoint. **Classifying, organising, constructing**

2. Using a program such as Google Earth, try to locate any green roofs in Melbourne. How many green roofs can you find in the central business district? **Comparing and contrasting**

3. Use the internet to research Melbourne's award-winning building Council House 2.
   (a) What features of the building are *sustainable*?
   (b) Would you like to work in a building like this? Why or why not?
   (c) Should any of the *sustainable* features of this building be made mandatory for future building developments? Explain your response.
   (d) Outline any future plans for the building. Why are they being considered?
   Evaluating, predicting, proposing

4. Research the ways in which your local council is working at a local level to improve *sustainability*. Most councils have a section on their website dedicated to actions for *sustainability*. Work in a small group to create a short presentation on the various programs at work. What kind of programs can individuals participate in? **Classifying, organising, constructing**

5. Use the internet to find out how an existing building can be made more *sustainable.*

Examining, analysing, interpreting

6. As you get older, your needs, wants and priorities will *change*. Imagine you have now completed Year 12 and are ready to move out into your first share house. In a small group (representing your new housemates), agree on a list of 10 ways that you and your housemates could live more *sustainably*.

Evaluating, predicting, proposing

9.9 EXERCISES

Geographical skills key: GS1 Remembering and understanding **GS2** Describing and explaining **GS3** Comparing and contrasting **GS4** Classifying, organising, constructing **GS5** Examining, analysing, interpreting **GS6** Evaluating, predicting, proposing

9.9 Exercise 1: Check your understanding

1. **GS1** What type of dwelling is your residence?
2. **GS2** Explain why the types of households are going to *change* in the next 20 years in Australia.
3. **GS1** Who are the three key groups making our urban areas more *sustainable*?
4. **GS1** List the benefits of green roofs.
5. **GS2** Many governments offer subsidies when elements such as green roofs and solar panels are included in building designs. Why do you think they would do that?

9.9 Exercise 2: Apply your understanding

1. **GS5** How are Australian households predicted to *change* over the next 20 years? What type of household do you live in?
2. **GS6** As a teenager, what do you think are some of the advantages and disadvantages of living in a high-rise or apartment building?
3. **GS6** Study **FIGURES 3**, **4**, **5** and **6**.
   (a) What are some ways in which governments can make changes to create a more liveable future?
   (b) What are some ways in which you, as a high-school student, can make changes to create a more liveable future?
4. **GS6** Make your own personal *sustainability* plan, using a mind map to help categorise your ideas. Consider how you could make *changes* in various areas of your life (school, home, sport, hobbies). List the actions that you would take, and identify what the outcome would be. For example, 'I could ride to soccer practice after school instead of being driven'. Outcome: reduced GHGE from the family car.
5. **GS6** Propose simple actions that you and your classmates could do to improve the *sustainability* of your school. Explain how each of these actions can improve *sustainability* as well as improving the quality of life for your school's students.

Try these questions in learnON for instant, corrective feedback. Go to www.jacplus.com.au.

9.10 Thinking Big research project: Electric vehicle report

SCENARIO

Your local council has asked you to prepare a report on the viability of electric vehicle (EV) use in your suburb. You are to research and assess electric vehicle use and availability and compare it to petrol and diesel alternatives. Your detailed report outlining the viability of EVs in your suburb or town will be presented at the next council meeting.

Select your learnON format to access:

- the full project scenario
- details of the project task
- resources to guide your project work
- an assessment rubric.

 Resources

💡 **projectsPLUS** Thinking Big research project: Electric vehicle report (pro-0175)

9.11 Review

9.11.1 Key knowledge summary

Use this dot point summary to review the content covered in this topic.

9.11.2 Reflection

Reflect on your learning using the activities and resources provided.

 Resources

🎞 **eWorkbook** Reflection (doc-31358)

　　　　　Crossword (doc-31359)

 Interactivity Managing and planning Australia's urban future crossword (int-7602)

KEY TERMS

affordability the quality of being affordable — priced so that people can buy an item without inconvenience

bi-articulated bus an extension of an articulated bus, with three passenger sections instead of two

congestion the state of being overfilled or overcrowded

development corridor area set aside for urban growth or development

ecological footprint the amount of productive land needed on average by each person in a selected area for food, water, transport, housing and waste management

family household two or more persons, one of whom is at least 15 years of age, who are related by blood, marriage (registered or de facto), adoption, step-relationship or fostering

incentive something that motivates or encourages a person to do something

infrastructure the facilities, services and installations needed for a society to function, such as transportation and communications systems, water and power lines

livelihood job or skill that supports a person's existence, so that they can have the necessities of life

rebate a partial refund on something that has been bought or paid for

retrofitting adding a component or accessory to something that did not have it when it was originally built or manufactured

triple bottom line an accounting term for measuring the success of a city, country or organisation by the health of its environment, its society and its economy

urban relating to a city or town; the definition of an urban area varies from one country to another depending on population size and density

urban sprawl the spreading of urban areas into surrounding rural areas to accommodate an expanding population

viable capable of working successfully

GLOSSARY

affordability the quality of being affordable — priced so that people can buy an item without inconvenience

altitude height above sea level

aquifer a body of permeable rock below the Earth's surface that contains water, known as groundwater

archaeological concerning the study of past civilisations and cultures by examining the evidence left behind, such as graves, tools, weapons, buildings and pottery

avalanche a sudden downhill movement of material, especially snow and ice

backwash the movement of water from a broken wave as it runs down a beach returning to the ocean

barge a long flat-bottomed boat used for transporting goods

bi-articulated bus an extension of an articulated bus, with three passenger sections instead of two

blizzard a strong and very cold wind containing particles of ice and snow that have been whipped up from the ground

catchment area of land that drains into a river

clearfelling a forestry practice in which most or all trees and forested areas are cut down

clinometer an instrument used for measuring the angle or elevation of slopes

compost a mixture of various types of decaying organic matter such as dung and dead leaves

congestion the state of being overfilled or overcrowded

convection current a current created when a fluid is heated, making it less dense, and causing it to rise through surrounding fluid and to sink if it is cooled; a steady source of heat can start a continuous current flow

converging plates a tectonic boundary where two plates are moving towards each other

coral atoll a coral reef that partially or completely encircles a lagoon

cultural relating to the ideas, customs and social behaviour of a society

deposition the laying down of material carried by rivers, wind, ice and ocean currents or waves

destructive wave a large powerful storm wave that has a strong backwash

development corridor area set aside for urban growth or development

divergent plates a tectonic boundary where two plates are moving away from each other and new continental crust is forming from magma that rises to the Earth's surface between the two

downstream nearer the mouth of a river, or going in the same direction as the current

drainage basin an area of land that feeds a river with water; or the whole area of land drained by a river and its tributaries

ecological footprint the amount of productive land needed on average by each person in a selected area for food, water, transport, housing and waste management

ecosystem an interconnected community of plants, animals and other organisms that depend on each other and on the non-living things in their environment

ecotourist a tourist who travels to threatened ecosystems in order to help preserve them

epicentre the point on the Earth's surface directly above the focus of an earthquake

erosion the wearing away and removal of soil and rock by natural elements, such as wind and water, and by human activity

escarpment a steep slope or long cliff formed by erosion or vertical movement of the Earth's crust along a fault line

estuary the wide part of a river at the place where it joins the sea

ethnic minority a group that has different national or cultural traditions from the main population

evapotranspiration the process by which water is transferred to the atmosphere from surfaces such as the soil and plants

family household two or more persons, one of whom is at least 15 years of age, who are related by blood, marriage (registered or de facto), adoption, step-relationship or fostering

fault an area on the Earth's surface that has a fracture, along which the rocks have been displaced

field sketch a diagram with geographical features labelled or annotated

flash flood a flood that occurs very quickly, often without advance warning

floodplain an area of low-lying ground adjacent to a river, formed mainly of river sediments and subject to flooding

fly-in, fly-out (FIFO) a system in which workers fly to work, in places such as remote mines, and after a week or more fly back to their home elsewhere

focus the point where the sudden movement of an earthquake begins

food miles the distance food is transported from the time it is produced until it reaches the consumer

geographical factors reasons for spatial patterns, including patterns noticeable in the landscape, topography, climate and population

geothermal energy energy derived from the heat in the Earth's interior

glacier a large body of ice, formed by an accumulation of snow, which flows downhill under the pressure of its own weight

gorge narrow valley with steep rocky walls

groundwater water that seeps into soil and gaps in rocks

habitat the total environment where a particular plant or animal lives, including shelter, access to food and water, and all of the right conditions for breeding

hard engineering a coastal management technique that involves using physical structures to control the effects of natural processes

high-density housing residential developments with more than 50 dwellings per hectare

host an organism that supports another organism

hotspot an area on the Earth's surface where the crust is quite thin, and volcanic activity can sometimes occur, even though it is not at a plate margin

human features structures built by people

humidity the amount of water vapour in the atmosphere

hunter–gatherers people who collect wild plants and hunt wild animals rather than obtaining their food by growing crops or keeping domestic livestock

hydroelectric dam a dam that harnesses the energy of falling or flowing water to generate electricity

ice ages historical periods during which the Earth is colder, glaciers and ice sheets expand and sea levels fall

incentive something that motivates or encourages a person to do something

indigenous peoples the descendants of those who inhabited a country or region before people of different cultures or ethnic origins colonised the area

indigenous native to or belonging to a particular region or country

infrastructure the facilities, services and installations needed for a society to function, such as transportation and communications systems, water and power lines

intermittent describes a stream that does not always flow

intermittent creek a creek that flows for only part of the year following rainfall

islet a very small island

katabatic wind very strong winds that blow downhill

lagoon a shallow body of water separated by islands or reefs from a larger body of water, such as a sea

landslide a rapid movement of rocks, soil and vegetation down a slope, sometimes caused by an earthquake or by excessive rain

leaching a process that occurs in areas of high rainfall, where water runs through the soil, dissolving minerals and carrying them into the subsoil. The process can be compared to a coffee pot in which the water drips through the coffee grounds.

liquefaction transformation of soil into a fluid, which occurs when vibrations created by an earthquake, or water pressure in a soil mass, cause the soil particles to lose contact with one another and become unstable; for this to happen, the spaces between soil particles must be saturated or near saturated

lithosphere the crust and upper mantle of the Earth

livelihood job or skill that supports a person's existence, so that they can have the necessities of life

longshore drift a process by which material is moved along a beach in the same direction as the prevailing wind

low-density housing residential developments with around 12–15 dwellings per hectare, usually located in outer suburbs

mantle the layer of the Earth between the crust and the core

meander a winding curve or bend in a river

medium-density housing residential developments with around 20–50 dwellings per hectare

megacity city with more than 10 million inhabitants

megaregion area where two or more megacities become connected as increasing numbers of towns and ghettos develop between them

metropolitan region an urban area that consists of the inner urban zone and the surrounding built-up area and outer commuter zones of a city

microclimate specific atmospheric conditions within a small area

migrant a person who leaves their own country to go and live in another

migration the movement of people (or animals) from one location to another

moraine rocks of all shapes and sizes carried by a glacier

nomadic describes a group that moves from place to place depending on the food supply, or pastures for animals

Pangaea the name given to all the landmass of the Earth before it split into Laurasia and Gondwana

peninsula land jutting out into the sea

per capita income average income per person; calculated as a country's total income (earned by all people) divided by the number of people in the country

perennial describes a stream that flows all year

permafrost a layer beneath the surface of the soil where the ground is permanently frozen

physical process continuing and naturally occurring actions such as wind and rain

plateau an extensive area of flat land that is higher than the land around it. Plateaus are sometimes referred to as tablelands.

population density the number of people living within one square kilometre of land; it identifies the intensity of land use or how crowded a place is

population distribution the pattern of where people live; population distribution is not even — cities have high population densities and remote places such as deserts usually have low population densities

prevailing wind the main direction from which the wind blows

pull factor favourable quality or attribute that attracts people to a particular location

push factor unfavourable quality or attribute of a person's current location that drives them to move elsewhere

rain shadow the drier side of a mountain range, cut off from rain-bearing winds

rebate a partial refund on something that has been bought or paid for

retrofitting adding a component or accessory to something that did not have it when it was originally built or manufactured

rift zone a large area of the Earth in which plates of the Earth's crust are moving away from each other, forming an extensive system of fractures and faults

river delta a landform created by deposition of sediment that is carried by a river as the flow leaves its mouth and enters slower-moving or stagnant water. Can take three main shapes: fan shaped, arrow shaped and bird-foot shaped.

sanitation facilities provided to remove waste such as sewage and household or business rubbish

sastrugi parallel wave-like ridges caused by winds on the surface of hard snow, especially in polar regions

sea change movement of people from major cities to live near the coast to achieve a change of lifestyle

sediment material carried by water

seismic waves waves of energy that travel through the Earth as a result of an earthquake, explosion or volcanic eruption

selective logging a forestry practice in which only selected trees are cut down

shell middens Indigenous archaeological sites where the debris associated with eating shellfish and similarfoods has accumulated over time

shifting agriculture process of moving gardens or crops every couple of years because the soils are too poor to support repeated sowing

slum a run-down area of a city characterised by poor housing and poverty

soft engineering a coastal management technique where the natural environment is used to help reduce coastal erosion and river flooding

soluble able to be dissolved in water

species a biological group of individuals having the same common characteristics and being able to breed with each other

stalactite a feature made of minerals, which forms from the ceiling of limestone caves, like an icicle. They are formed when water containing dissolved limestone drips from the roof of a cave, leaving a small amount of calcium carbonate behind.

stalagmite a feature made of minerals found on the floor of limestone caves. They are formed when water containing dissolved limestone deposits on the cave floor and builds up.

subsistence producing only enough crops and raising only enough animals to feed yourself and your family or community

sustainable development economic development that causes a minimum of environmental damage, thereby protecting the interest of future generations

swash the movement of water in a wave as it breaks onto a beach

tectonic plate one of the slow-moving plates that make up the Earth's crust. Volcanoes and earthquakes often occur at the edges of plates.

temperate describes the relatively mild climate experienced in the zones between the tropics and the polar circles

transportation the movement of eroded materials to a new location by elements such as wind and water

treaty a formal agreement between two or more countries

tree change movement of people from major cities to live near the forest to achieve a change of lifestyle

tributary a river or stream that flows into a larger river or lake

triple bottom line an accounting term for measuring the success of a city, country or organisation by the health of its environment, its society and its economy

urban sprawl the spreading of urban areas into surrounding rural areas to accommodate an expanding population

urban relating to a city or town; the definition of an urban area varies from one country to another depending on population size and density

urbanisation the growth and expansion of urban areas

utilities services provided to a population, such as water, natural gas, electricity and communication facilities

viable capable of working successfully

volcanic loam a volcanic soil composed mostly of basalt, which has developed a crumbly mixture

watershed an area or ridge of land that separates waters flowing to different rivers, basins or seas

weathering the breaking down of bare rock (mainly by water freezing and cooling as a result of temperature change) and the effects of climate

INDEX

Note: Figures and tables are indicated by italic *f* and *t*, respectively, following the page reference.

Nepal   2015 earthquake   124–5
New York City   236, 237, 239–41
New Zealand   33, 34, 102
   Christchurch earthquake (2011)
      125–6
   Franz Josef Glacier in   72f
   liquefied soil in Christchurch
      132f
   Mount Taranaki   138–9
nomadic   97, 164
North American Plate   127
Nullarbor Plain   26

O

ocean trenches   127
oceanic islands   15
orangutans, deforestation and   170–2
organic compost   153
Ota, Japan   246–7
oxbow lake   67

P

Pacific Ring of Fire   34
Pangaea   114, 149
Pearl River delta (PRD)   227–8
Pedirka Desert   93–4
Penan people of Malaysian Borneo
   164–5
peninsula   57, 77
per capita income   220
perennial river   65, 77
permafrost   15, 46
photographs, describing   224
physical process   56, 77
pictographs   205
place, as geographical concept   5–6
plateau   15, 46, 236
plateau mountains   113–14
playas   88, 90–1
polar deserts   85, 99
polar regions   15
pollution
   air pollution   197, 220, 222,
      233, 237
   transport and   220
pollution density, disease and   197
population density   213
   Australia   184–7
   average for each continent   187
   definition   184
   disease and   197
population distribution   213
   Australia   6f, 184–7
   definition   184
population growth, Australia   203f
population profiles   212

PRD   *see* Pearl River delta
prevailing wind   54, 77
public transport
   benefits   270–2
pull factors, rural–urban migration
   208, 215–6, 256
push factors, rural–urban migration
   208, 215

Q

quality of life   257, 258, 280, 281

R

rain-shadow deserts   83–4
rainforests   15
   Amazon rainforest   158–60
   in Australia   156–7
   benefits of   160
   causes of deforestation   167–9
   conservation   174–8
   definition   151
   deforestation   167–9
   ecosystems   152–5
   impacts of   169
   indigenous people and   163–7
   location   151
   physical processes   152
   tropical rainforests   151–2
   types   151f
Ranger uranium mine   40
rebates   280, 284
Regional Sponsored Migration
   Scheme (RSMS)   204, 204f
resources, coastal   59
retrofitting   280, 284
Richter scale   124, 126
rift zones   136, 149
river
   cross-section   65
   formation   65
   long profile   67
   longest   65
   meandering   67
   mouth   67
   upper course   66–7
river deltas   67, 77
river landscape
   formation of   65–8
   management of   69–72
river systems and features   65–8
Rocky Mountains, North America
   18, 117
RSMS   *see* Regional Sponsored
   Migration Scheme
rural lifestyle   261–3

rural–urban fringe,
   housing   264
rural–urban migration
   in China   208
   in India   216–8
   pull factors   208, 215–6, 256
   push factors   215

S

sacred places, in mountain landscapes
   121–2
Sahara Desert   83, 100
saltpans   88, 90
sand dunes   88–92, 96
   formation of   53f
sanitation   197, 220, 224, 236
sastrugi   101
savanna   15
scale
   as geographical concept   10–2
   using to calculate distance   94
SDGs   *see* Sustainable Development
   Goals
sea change population movement
   206, 213
seasonal agricultural workers   206–8
sediment   28, 46
seismic waves   124, 149
selective logging   167
SEZ   *see* Special Economic Zone
shell middens   59, 77
shield volcanoes   141, 141f
shifting agriculture   163
Sierra Nevada Range   115f, 114–16
Simpson Desert   85, 89, 92
sketch map   255
slums   219–21, 222f, 228, 236
soft engineering   56, 77
soil   20–2
   Australian   20
   composition of   20
   formation of   21–2, 24
   profile   22–4
solar panels   278, 280, 281
soluble bedrock   25, 46
South America, cities   226
South Australia   31
space, as geographical concept   4–5
Special Economic Zone (SEZ)   227
species, diversity   150
SPICESS (geographical
   concepts)   4
Sri Lanka, urban greening program
   245–6
stalactites   24, 46
stalagmites   24, 46